Best Ideas
For ORGANIC
Vegetable Growing

Best Ideas
For ORGANIC
Vegetable Growing

by the staff of
ORGANIC GARDENING
AND
FARMING MAGAZINE

J. I. Rodale, *Editor-in-Chief*

Robert Rodale, *Editor*

Jerome Olds, *Executive Editor*

M. C. Goldman, *Managing Editor*

Maurice Franz, *Managing Editor*

Edited By

Glenn F. Johns

RODALE BOOKS, INC.
Emmaus, Penna. 18049

Distributed by David McKay Co., Inc.

ISBN 0-87857-042-X hardcover

LIBRARY OF CONGRESS CATALOG CARD NUMBER 77-90877

COPYRIGHT 1969 BY RODALE BOOKS, INC.

ALL RIGHTS RESERVED

PRINTED IN UNITED STATES

16   18   20   19   17   Hardcover

CONTENTS

Introduction

We have prepared this book to bring you the best ideas on growing vegetables in your home garden. Organic gardeners throughout the United States as well as in different parts of the world have used the techniques you're about to read to harvest vegetables rich in nutrients and flavor.

Within the sections for each vegetable, you'll learn about planning, seed starting, soil conditioning, varieties, insect and disease protection, harvesting tips—as discovered and developed by many of the most successful organic gardeners.

Discover what mulches are best for tomatoes . . . soil conditioners that really work for corn plantings . . . ideas for a second harvest . . . methods for effective weed control . . . when no-cultivating is most effective . . . how to keep disease and insect damage to a minimum without using any poison sprays.

The information comes from gardeners with all types of soil problems and climatic variables. Whether your vegetable plot is measured in square feet or acres, we believe you will use these ideas to grow more organic vegetables.

⊗ ⊗ ⊗

Growing vegetables has always been the most significant part—the *raison d'etre*—of the organic garden. There's no better way to get the quantities of high-quality food you and your family want.

In compiling BEST IDEAS FOR ORGANIC VEGETABLE GROWING, the editors carefully examined thousands of reports by organic vegetable growers. The most popular vegetables have generally been given the most space. Always it has been the idea which worked that prompted the write-up to be included in this book.

We hope—and believe—the ideas you are about to read will add to the success of your own vegetable garden.

ASPARAGUS

Starting the Easy Way

Despite the fact that for generations the "experts" have continued to insist that digging an 18- to 24-inch-deep trench is the essential first step in starting an asparagus bed, I can assure you that it isn't necessary at all. I also disagree with them on the desirability of planting two-year-old roots. Having grown asparagus for nigh on to 50 years, I think I can claim to be something of an expert myself.

First, I decided to raise my own plants from seed and to transplant when they were a year old, before their root systems spread out too far. This would enable me to move each plant with practically all the roots intact.

Accordingly, I planted the seeds sufficiently far apart to give me ample digging space around each, and I also prepared the future bed by simply spading into it all the composted material I was able to turn under. The following spring, I transplanted the strongest plants and threw away the rest.

At least, I thought I'd disposed of all of the rejects; but I later discovered that I had overlooked two, possibly because they were too small, at the time, to attract my attention. However, by that time they appeared to be growing so well that I decided to leave them where they were and see what happened.

Thus, I had asparagus growing by 3 different planting methods: A—the original trench-style bed planted 4 years previously with two-year old roots; B—an experimental bed, planted with one-year old roots, but without a trench; and C—two plants, grown from seed and never moved from their original location.

Of course, group A, having several years head start, was producing quite well by this time, but the rate of growth of the other plants was very rapid, and by the end of that year they had reached a height of from 5 to 6 feet and I was able to make light cuttings the following spring, two years after planting the seed.

That was 18 years ago, and during the intervening years it became increasingly obvious that my "trenchless," seed-grown plants were far superior to those set out by traditional methods. Last year I was forced to dig up the original bed as the stalks produced had dwindled to pencil thickness, or smaller. Groups B and C still produce stalks as thick as my thumb and show no sign yet of giving out, in spite of the fact they are growing in about 8 inches of topsoil with hardpan immediately below.

From this experience, I'm convinced that the most difficult and least productive way to grow asparagus is by the long-recommended trench-digging method, using two-year old roots.

A much easier and more productive plan is to grow your own plants from seed and transplant them the following spring, while it's still possible to move them with practically all roots intact in a ball of earth.

But by far the easiest and best way of all is to plant the seed right in the permanent bed and never move the plants at all. Don't bother with a trench, but simply spade all of the organic plant food you can work into the top 10 inches of soil where you plan to have the bed. Do this during the summer or fall prior to planting the seed, as this permits the organic matter to become incorporated with the soil and it eliminates air pockets. A good dressing of lime is also recommended.

This is the only "work" you'll have to do to produce the bed, but it's no more than should be done in any garden prior to planting anything. Remember, too, it's the last time you'll have to do any spading for the next 18 years or more.

Early the following spring, plant one seed about every 6-8 inches along the row, keeping the rows about 3 feet apart. The following spring, after the first growing season, allow the plants to start up sufficiently to enable you to identify the strongest plants and then pull up, or dig up, all the rest and discard them, leaving your permanent plants an average of about 2½ feet apart. That's all there is to it.

In selecting which plants to leave and which to eliminate (if you grow them from seed in the permanent bed), there invariably will be some fine plants too close to others to allow them to remain. Such plants, being only a year old, can be dug up easily and replanted in any open places you happen to have along the back or side fences. They provide a decorative background and they'll also supply a few additional cuttings if the soil is reasonably fertile, but by all means resist the temptation to leave too many plants too close together. The crowns

Many gardeners plant asparagus in trenches, working compost into the soil with a tiller.

Crowns of one-year plants are set close to surface at left, while compost mixed with manure is being spread in trench at right.

of asparagus plants spread year after year, and if each plant isn't given sufficient feeding area and must compete with other plants for plant food, you defeat your objective, which is to have thick, healthy stalks.

—Holcomb York

A Lifetime of Good Eating

Asparagus is best planted early in spring, as soon as the ground can be worked. It isn't necessary to cultivate the entire bed, but the plants should be kept moist while it's being prepared. Mark off the rows, and dig a trench 15 inches deep and 18 inches wide along their full length. Firm down a 4-inch mixture of half well-rotted manure and half aged compost on the bottom, and top it with 4 inches of good topsoil.

Using only strong, healthy plants spaced 20 inches apart—close planting interferes with good production—set the heart of each on a small mound of soil so its roots radiate out—like a tree's. Cover with just two inches of fertile topsoil worked in carefully around each plant with the fingers and firmed down gently so the young roots are not harmed. To avoid plant loss from suffocation, don't fill the trench completely, and be sure to water it well when you're done planting.

Thin young shoots will soon appear and your asparagus bed will be on its way. By fall the trench will be filled in again if you throw about one inch of soil in around the young plants every time you hoe down the weeds. If drought hits, flood the trenches once

7

each two weeks.

Start mulching your young plants the second spring, after you have applied a two-inch layer of old manure or compost over the whole bed. Mulch the entire length to keep it weed-free using available organic materials—old hay, leaves, straw, salt hay and dried grass clippings—about 8 inches for a season. If you want a steady yearly supply of thick, delicious spears, repeat this practice every spring.

Do not cut shoots from a bed of one-year-old plants until the third year, and then take just 3 cuttings. The young plants are still forming productive crowns beneath the surface, and need plenty of foliage to produce growing energy.

You may harvest shoots either by cutting them off two inches down with a sharp knife, or by snapping them off at surface level without injuring the tender shoots that form just below the ground. If you must use a knife, slip it into the earth next to the shoot, twist the wrist slightly, and the shoot will fall over without damaging nearby spears. For blanched asparagus shoots, hill each row slightly early in spring.

Do not put your asparagus bed to sleep until after heavy frosts have browned the foliage, at which time the stalks may be cut at ground level and laid across the bed to hold the winter mulch in place. When stalks are loaded with red seeds, it is wiser to cut them down, adding them to the compost pile to avoid a multitude of wild seedlings next May.

The one thing to remember is that

Plant's roots should radiate out from mound. Cover with two inches of soil, and firm into place gently without hurting young roots.

Cut asparagus shoots when 4 to 8 inches high, before the scales on the tips begin to open.

asparagus must be fed and mulched with organic matter every spring if it is going to produce an abundance of thick spears. The more black soil that is formed over the crowns—the more spears you will harvest as the years go by.

—Betty Brinhart

Don't Work So Hard for Asparagus!

I have two 50-foot rows of asparagus, one of which was planted 30 years ago in the old-fashioned way, before I knew better; that is, in a very deep, wide trench. The other was put in two years later, in a shallow trench. The two rows are doing equally well.

For the first 14 of my gardening years I covered my asparagus each fall with manure, cultivated it each spring, weeded it all summer long.

Then, one fine day in April, I got the bright idea of abandoning plowing and hoeing and weeding. I covered my plot with hay, left it there, added more now and then, and for the past 16 years the work in my garden has consisted of replenishing the mulch here and there, planting, and picking my wonderful produce.

This of course includes the asparagus bed, and, as I said above, I can't for the life of me figure out why anyone should think that the hay should be removed in the spring and the bed cultivated. The tips will come up right through the hay, so why disturb it?

Or you might push the mulch back on only a part of your bed, which would give you a longer season. You would in this way be cutting one section a week or ɩ.ɾo earlier than the

9

rest, and you could cut the second section a week or so after you've stopped cutting the first.

In general, leaves are a good mulch, but loose hay is the best for asparagus; leaves, or hay that has been baled, may prevent the sprouts from coming through. Straw is all right. However, J. A. Eliot of New Jersey, an asparagus expert, believes that hay is the best mulch of all; he says that for nutritive value it is superior to manure. And his reasoning is that part of the nutrients in hay, which is fed to horses and cows, go to build up the body of the animals and to make milk; manure is the residue. But a rotting hay mulch still has all the nutrients left in it.

For the past 16 years I have used no fertilizer of any kind on any part of my garden except the rotting mulch and cottonseed meal. I broadcast the latter in the winter, at the rate of 5 pounds to every 100 square feet of my plot. I'm not really convinced that my soil needs the meal, but I have been told that it does, for nitrogen.

Now a word about picking asparagus. People can't seem to get away from that slow business of cutting it with a sharp knife, or a two-pronged asparagus cutter, just below the surface of the ground. For my money, that method has 4 things wrong with it: it takes quite a little time; one is likely to injure a nearby shoot which doesn't yet show above the ground; the stalk-ends are dirty; and the tough part has to be cut off and disposed of.

My system is much simpler: I walk down the row and snap off any stalk which has matured, and since I break it where it's tender, there's nothing to be cut off afterward. And the stalks are so clean that all they need is a quick rinse under cold running water.

The amount of money my method of growing this vegetable will save you depends on how much you have to pay for hay, and how much you would spend for fertilizer if you grew asparagus the old-fashioned way. But I am sure my system will save you a tremendous amount of time and energy.

—Ruth Stout

Get the Most from Your Asparagus

For maximum length of edible stalk pick when the "bird shot" has opened but before stalk has started to branch out. Store customers, of course, want the closed or folded head. The best-tasting asparagus, though, are the fat, purplish-pink spears about 3 to 4 inches.

I break stalks off just above the butt. Since my plants are not deep-planted, I don't need to cut below the surface, possibly injuring other shoots just forming.

When the bed is young, the space between plants can be used for intercropping. Grow things like annual flowers, or Alpine strawberries or parsley. Don't plant radish—there's too much manure for it. Summer lettuce benefits from the partial shade. My bed is now too dense; the lettuce would suffer, though not the asparagus. But I could grow lettuce as a parallel edging two or three feet from the asparagus. Or plant a permanent little one-foot hedge of the evergreen herb, winter savory.

To get more plants at no cost, you can save your own seed. Pick the red, ripe berries on female plants. Spread them out in a protected, airy place and let dry. Keep over winter in a dry place. Soak overnight before planting the next spring. Or, pick ripe berries, soak to loosen shell. Mash carefully, then place in a sieve and wash out shell, leaving seeds. Dry in sunny place, then store.

Seed-saving makes an interesting experiment, but actually there are always plenty of volunteer seedlings to start a new bed with. Left in the old bed, they would turn it into a ferny jungle. Fall is the best time to transplant seedlings that have come up that season (from berries fallen the year before). Seed-sown plants from one's own stock may not run true, but they're usually good strong specimens.

—Ruth Tirrell

BEAN

No Bean Could Be Better

Of all the green beans we have grown — flat-podded, round-podded, Italian, horticultural, pole — we consider Black Valentine, an early bush bean, the most outstanding. Now we plant it year after year. This bean not only produces crisp, dark-green oval pods of fine flavor, it gives an unusually heavy yield.

About 4 years ago, when our soil lacked organic matter, we sowed two rows of bush beans before we had any finished compost. The resulting plants had small light-green leaves and slender stems. As soon as our compost could be used, we mulched the beans heavily with it. Within a week the plants looked noticeably healthier, and soon after we were astonished to see the plants had grown strong and stocky, with their leaves a deep green.

Before this time we grew a different green bean each spring. Although each was satisfactory, we felt something was lacking. There must be a better bean, we thought, as we looked forward to trying still another kind next year. Oddly, we could find little difference between some varieties. Then we found Black Valentine, and it was just what we had been looking for.

Bush beans should be sowed when the ground is fairly warm, after May first here in southern Massachusetts.

When we have planted too early, up to a third of the seeds have rotted in the cold, too-moist earth. Then there is the risk of late frosts, although if one is expected we cover the rows with apple and grape boxes or peach baskets.

Make a shallow depression with a metal rake along the row to be planted, and fill it with plenty of compost. After an inch or so of good loam is spread over it we have a long, low mound. That's an advantage, for our garden is low, bordering a drained meadow. Besides, the row will settle somewhat before the seedlings appear. We sow the beans in staggered double rows, 6 inches apart each way, and press them about 1½ inches into the earth. We often make the rows less than 18 inches apart, though it is preferable to space them wider so that second or third crops of other vegetables can be sowed between rows even before the beans stop bearing. If necessary, the stripped bean plants can be left awhile to protect or shade tender seedlings from the summer sun.

When the beans have two pairs of leaves we mulch with hay or grass clippings to conserve moisture and make weeding unnecessary.

It is surprising how many pickings can be taken from a row of Black Valentines. After picking two or three times a week for several weeks, we thought it about time to collect the remaining pods and yank out the plants to augment the compost pile. However, we were astonished to find that they had flowered again and another crop was ripening.

We remove the beans with a short sharp knife, cutting each pod just before the stem connection, which leaves the little hard nub on the plant. Cut-ting there with a knife or your thumbnail is easy on the plant and much time is saved in the kitchen; after rinsing, the pods need only to be lined up on the slicing board and cut across with a few quick slashes of a straight knife.

—Devon Reay

Water Bags As Mulch!

Plastic bags filled with water make a good mulch, says the USDA. String bean yields on an experimental plot in Idaho were improved by 20 percent after water-filled plastic bags were placed on the soil as a mulch. Most vegetables grow best within a fairly narrow range of soil temperatures, notes Sidney A. Bowers, ARS scientist conducting the research. Testing his water-mulch theory first on uncropped soil, he found the bags reduced maximum temperature and increased minimum temperature in the upper half-inch of topsoil. Mulched and check-comparison plantings of two heat-sensitive crops—corn and beans—showed as much as 12,000 pounds per acre difference in yields. Water mulches would allow gardeners to plant early in the spring with less risk of frost damage, Bowers believes, and should be especially helpful in regions where temperatures vary sharply during the growing season.

Royalty Beans—Fit for a King

Purple-podded beans, otherwise known as Royalty variety bush beans, are good eating for humans, but the Mexican bean beetle doesn't care for them. This makes them especially attractive to organic gardeners who are always on the lookout for natural methods of insect control.

Royalty beans will germinate in

cooler soils than most others. I like to plant one short row around the first of May. Here in northern Ohio the regular planting time is around Decoration Day. When frost threatens, I protect the seedlings with 3 or 4 thicknesses of newspaper set over the plant row in tent style and secured with stones. The purple beans are heavy producers, so that one short row gives us all we can eat until the main crop of regular varieties comes in. I then pull up the vines and plant something else in their place.

Round, tender and stringless, Royalty beans are tasty either fresh or canned — and freeze well, too. A bush-bean type, they're grown just as you would any other. Snap beans will do well on a soil too poor for other crops, but good garden soil will bring increased yields. When fertilizing, avoid excessive nitrogen. If your soil is reasonably fertile, a good dusting of ground phosphate rock for phosphorus and granite dust for potassium, added as you prepare the soil, is all that is necessary.

Once the beans are up well, I put a heavy mulch—hay in my case, though

Royalty beans germinate in cooler soils, are more insect-resistant than other varieties.

other things would do as well—along both sides of the rows. It's been my experience that this mulch does more to increase production by conserving moisture in the ground than anything else I do. Last year I had two very sad-looking rows of beans simply because I didn't get around to mulching them, and we had a dry summer. All the rest of my beans produced huge crops—far more than we could use.

Snap beans, whether green or purple, are one of our most important vegetables. Not only are they popular, but both the calorie and carbohydrate count are so low that dieters or those who must watch their sugar intake can eat all they want. Nutritionally, beans are a good source of vitamins A, C and B-complex, as well as iron and potassium.

—Lucille Shade

Make Your Pole Beans Grow TALL

When I used "regular" chemical fertilizers last year my pole beans refused to climb their wooden poles. So, I changed my strategy. I set 3 plants to each hill, spaced 4 feet apart and, for good measure, stashed away two good-sized shovels of manure and 4 shovels of rotted sawdust and hay into each hill.

I also used the same poles that I had used the year before. But before I put the poles up, I wrapped them with old pig wire fencing that had been stretched, bent and torn by the pigs into worthless junk, ready to be hauled away! The wire made 8-inch cylinders around the poles, and after it was bent down and fastened with more wire, it provided good support.

The beans came up quickly and soon began to climb the poles. I guided the new growth through the

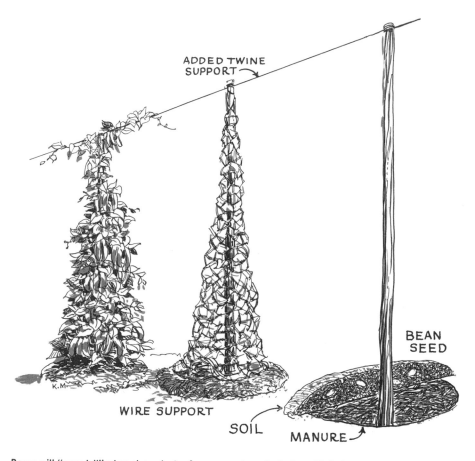

ADDED TWINE
SUPPORT

BEAN
SEED

WIRE SUPPORT

SOIL

MANURE

Beans will "grow tall" when given plenty of manure and sawdust, plus added wire support.

wire when they really began to intertwine with their lush growth. By the first of July the wires were completely covered, and soon the 8-foot poles were completely hidden by the large leaves, flowers and beans. But these vigorous vines kept right on growing, so I had to join the poles together with strands of baling twine and direct the growth along them. This I did, and then to keep the plants well supplied with water and nutrients, I put a thick mulch on the rows and began to make sheet compost between them. By fall I had 6 inches of the best compost, which was added to the large fall bulb plantings.

The beans from these plants were of the best quality, and in such abundance that many had to be added to the compost pile, in addition to those that were canned.

—John Francis

Pole Your Beans for Flavor

Here's how we plant our Blue Lake pole beans: First we get two 8-foot wood posts, 4 to 5 inches in diameter, and set them 18 inches deep at each end of the garden where our 50-foot bean row is planned. Fill holes and tamp well. At 5 feet from ground level, string tight barbed wire from one post to the other. Then brace post to the other. Then brace posts against the weight of the wire and beanpoles. `

Using the wire as a guideline, make a furrow two feet from the posts on each side down the full row. Next, place a 7-foot beanpole at every third barb along the wire, slanting one end in the furrow and the other against the barb. Place another from the opposite direction in the same manner. Then tie the two together, against the barb, with baling twine or butcher's cord.

For beanpoles you can use scantling (from a sawmill), dead sassafras trees (quite satisfactory), trimmed saplings (from clearing wooded areas), or anything that looks like a pole for the vines to climb on. If you have bamboo, that is fine, but keep poles to the 7-foot length.

Now, plant your beans — one on each side of each pole if you are optimistic, two if you are doubtful. For a thick row, plant two or three in between. Cover the seed, and if you are in a heavy rain area—as we are—run ditches one foot away from the rows for drainage. We have found that pole beans do not like too rich a soil, or a wet, soggy one—but do like medium-rich soil. Either a light application of compost or well-rotted manure is sufficient.

We plant pole beans on Good Friday. The first rain after the beans are at the 4-leaf stage, we mulch under the wire between the rows. In order not to retard growth, we wait until the weather warms up—usually late May—to mulch the outside of the rows. By that time the beans are on their merry way up the poles.

A word of warning: Don't make your beanpoles over 7 feet long unless you are over 6 feet tall — otherwise, you need a stepladder to pick them. Blue Lake pole beans can grow to 10 or 12 feet high. On our 7-foot poles the vines grow to the top, then drape down, enabling a shorter person to pick the top beans easily.

—E. T. Ott

Keep 'Em Climbing, Blooming and Producing

Here at cactus ranch we do enough small-scale, intensive gardening to provide fruits and vegetables for the table and for storage shelves. Come spring, we plant Kentucky Wonder pole beans along the 6-foot wire fences around two 12-foot-diameter garden plots. Well, for years our beans had been blooming, and we'd pick the excellent harvests at prime ripeness, eating them fresh as well as canning scores of quarts—and then would yank out the tall-climbing vines after the one-crop flush. We'd never heard of anybody doing anything else.

But I sure came in for a surprise last year when visiting a nearby organic gardening friend in Malibu. Although it was already late summer, I saw her pole beans on a long-row wire trellis in good leaf, blossoming and producing lots of young beans.

"That's the third crop this year on those pole beans!" my neighbor exclaimed. And that is how I learned you can easily grow two and sometimes three crops a season on the same vines—not only of pole beans but peas and butter beans, too. Of course, the triple crop return depends partly on a favorable climate—such as Southern California bestows—and planting at proper times.

Back at the ranch, I had Kentucky Wonders planted in single-row style along the 6-foot fence surrounding my two small vegetable plots. The vines had been dropping their leaves for 10 days and seemed to be saying, "We've had it for this year. Pull us out."

Following my neighbor's advice, I went over the vines lightly, shaking and pulling off the dead foliage (all the leaves). Then I watered deeply and loosened the soil around the thick stalks entering the soil. Now I went to the chicken house and pigeon lofts (I raise squabs) and got a few bucketfuls of rich manure. Next I filled the buckets with water, stirred well, and left them to liquify.

The first step the following day was to take a small bucket and copiously feed the pole bean plants this liquid manure. I poured on all the ground would absorb. Now a wait of two days for the liquid fertilizer to soak down, and finally a deep, slow watering. Then I pulled back a thin mulch of oat straw around the rows—and sat back and waited.

It seemed only a day or two before the buds and leaves began "busting out all over." The heavily cropped first-round pole beans had been given a natural "shock treatment" and burst into new life. In the following days of sunlight they grew apace and were soon covered with thousands of leaves, tiny white blossoms—and then beans. I harvested a 60-percent second crop, ate some fresh, and canned another 15 quarts for our light but windy Southern California winter.

Based on my year's experience and the advice of my friendly neighbor, the way to get the largest production from pole beans (besides pole peas and limas) is to plant as early as you can get the seeds into the garden. Or, start your seed in flats indoors or in a greenhouse, and when planting time comes put 3-inch-tall, well advanced bean plants out in the garden rows. This gives you a big time jump—literally—and even in colder climates will give you two good crops annually.

I find that liquified chicken and pigeon manure is just right for any of the garden plants. Tomatoes, cucumbers and peppers thrive on it. Throughout the year all manure scrapings from the lofts and henhouse are scattered in the garden. It's also of interest to note that my white leghorn hens will not transgress over a 3-foot-high wire fence.

And as for pole beans—every organic gardener shoud grow 'em. They're delicious for eating fresh, very easy to can, and are an excellent source of vitamins, calcium and iron.

—Gordon L'Allemand

A "Leaf Sled" for Limas

Tired of making so many trips last fall to tote compacted dry leaves in our old 5-by-7 bedspread, I decided to build a large leaf sled out of a few spare pieces of stock wire. The "bottom"—5 by 12 feet—came from the 60-foot fence of pole lima beans which had stood 7 to 10 feet high. Then I found two 10-foot pieces, each 3 feet wide, for the sides and pieced them out with a 6-foot section that I used for the back. It took a little extra "weaving" and using binder twine, but in a little while I had a serviceable sled which slides easily over the ground and brings me plenty of leaves with very little effort.

—Edward P. Morris

Beating the Bean Beetle

Last year our bean crop was practically destroyed by bean beetles, even though we had a beautiful planting of marigolds with the beans. So this year we decided to plant beans in an entirely different location. However, shortly after the Fourth of July, I had a short vacant row of about 40 feet, and decided to plant it with bunch beans for a later picking. The latter part of August, I was picking some beans from this short row, when an elderly neighbor came past and asked if I was picking beans or bugs. Upon being advised that I was picking beans, he stated that they must have been planted after June 21, or the longest day of the year. After questioning him on this statement, he advised me that as a very small child he could remember hearing his father say that beans planted after the longest day of the year would not be bothered by bugs.

In early August we had some more space that was vacant and decided to plant some more beans, and this last planting has done very well. These plantings in July and August were in the same ground as those last year—only last year the planting was made in May and was destroyed by bean beetles. Both plantings made after June 21 this year have never been bothered with a single beetle.

If this theory proves to be a fact, we can certainly wait until the latter part of June or the first of July to plant our main bean crop.

—Charles W. Yergan

Lima Beans By the Carton

One winter I had sweet peas growing on the wire I normally use for my limas. They were still looking so pretty that I hated to pull them up when it was time to plant the limas. I solved the problem by starting the beans in discarded milk cartons, which I cut down to about 3 or 4

Fordhook 242 is a choice bush variety that matures large beans earlier than pole types.

inches, punching 4 holes, one on each side, for drainage.

Since my soil tends to heavy clay, I added sand to make it lighter and more porous, and then filled the cartons with this mixture, giving the beans a more friable soil than they would have if planted directly in the ground. The soil I use is average garden dirt with no compost added; too much richness would not be desirable since I give the beans at least 3 feedings besides mulching them heavily.

I plant two beans to each carton and, when they are a few inches tall with their second set of leaves, I discard the weaker plant in each container. The cartons are placed in a flat under a roof overhang so they are not rained upon. Don't overwater the newly sprouted beans, because the plants have not yet acquired their true leaves and rot very easily at this stage. Gardeners with only a small backyard planting of beans can gain a month or more by using cartons if their climate is too cold for an early spring planting.

I find I am less likely to overwater the cartons when I use a drinking glass.

Sometimes, when I am too rushed to start them in cartons, I dig a small trench, covering the bottom with a layer of sand, and then plant the beans directly on the sand. Any excess water drains off quickly and does not cause the young plants to rot. I am careful to water only when the ground feels dry, and then very sparingly, just enough to change the color of the soil.

When the bean plants are tall enough to start climbing on their supports, I set them out in the garden. After watering each one well, I take a pair of scissors and carefully cut off the sides and bottom of the carton. Very gently—the roots are not even slightly disturbed—I set each bean plant with its intact cubicle of soil into a hole prepared with my trowel. In my heavy soil the pole limas do quite well planted only two to three inches apart.

A couple of weeks after planting I mulch the bean row with a very thick layer of bagasse — shredded sugar cane stalks — hose the mulch down well with a fine spray, and then give each plant a cup of liquid manure so the freshly applied mulch will not steal nitrogen from the plants just when they need food. I use 3 teaspoons of cow manure to each gallon of water.

Even a small backyard planting of lima beans like this can be very worthwhile if your soil is reasonably good. From only 7 feet of row, the least I have ever harvested was about 10 pounds. One year I gathered 16 to 24 pounds—but that was an unusually good year.

One good thing about growing butter beans in the home garden is that they don't have to be eaten the same day they are picked. Sometimes you may get only a half pound or less at a picking. But if allowed to remain in their pods without being shelled, and placed in a grocery bag in the vegetable compartment of your refrigerator, they will keep up to a week by which time you will have enough to cook.

Ripe pods should never be allowed to remain on the vines because this will cause the plants to stop producing. If picking is done carefully on a weekly basis, the vines will continue to set flowers and pods until frost.

However, a great many pods may be permitted to vine-ripen before the plants are pulled at the end of the season. When completely dry, these organically-grown beans may be used either dried, or kept for seed. It is wise to put enough aside for next year's planting, because they produce a heavier and better crop in the following season.

After they have set their first crop, I again give the vines a feeding to encourage future production. Later on in the season I may occasionally give them another feeding. If I notice a great many yellow leaves suddenly appear, this is a sign that the vines need more nitrogen and should be fertilized.

While the slowly rotting mulch provides a little food during the season, it may have to be replenished if it weathers down too much. I put on about 3 or 4 inches of bagasse which rots very slowly, so I don't have to add more. However, other mulches may not last as long, so a new layer may have to be added. In regions where the soil is very light and sandy, more fertilizer or compost may be needed.

Here in the Deep South it rains a lot during the summer, so I usually don't have to water as the mulch keeps the ground moist. I do water more in early summer, when it is generally drier and the vines are not yet throwing much shade over the roots and stems of the plants.

If you can't spare enough space to accommodate a continuous planting of 6 or 7 feet, perhaps you can plant your beans on poles set here and there in your regular beds. I once had a hardware man cut some wire up into one-foot pieces for my beans to climb on. These long strips of wire were nailed to tall stakes which were staggered here and there between perennial and annual flowers. I didn't care much for the effect but it did prove to be a real space-stretcher. A lady whose yard was entirely paved planted her pole limas in large tubs and arranged wires for them to climb on. So if you really like limas and are determined to plant some, chances are you can find a way to fit them into your planting scheme even if you are cramped for space! You'll be glad you did.

—Phyllis Holloway

This Year, Grow the Unusual Beans

The Dixie Butterpea—it's actually a small kind of lima—is one of those rewarding but somehow overlooked leguminous vegetables that you just can't afford to ignore. Just like its hard-working cousins —the Long Pod broad bean, the Asparagus Pea and the plain peanut—it is a soil-builder that improves your homestead while giving you bumper crops of flavorful and vitamin B-laden food. So, make it a point to plant one or all of these rewarding plants this year, starting with the Dixie Butterpea.

It looks like a pea, is quite hardy, and may be grown in northern as well as southern gardens. Maturing more quickly than the larger limas, it is also a willing "volunteer" who comes up again the following spring when missed in the harvesting. Like its brethren, it may be cooked either green or dried, with the dried variety offering a distinctive, nutty flavor quite unlike any other bean.

Bush-type plants, butterpeas are vigorous and very prolific growers, yielding an abundance of small-to-medium-sized pods, each of which contains several butterpeas. They continue to produce until taken by frost, and are little troubled by garden pests. Besides the regular Dixie Butterpeas there is another variety, the speckled Dixie Butterpea, which is similar, but with speckled seeds and a somewhat more pronounced flavor.

In addition to being cooked green, matured, or dried, or used in soups, Dixie Butterpeas may be canned or frozen, and are very well adapted to any of these processes. The dried beans are excellent when baked.

Dixie Butterpeas are best planted with the eye down. As is the case with most garden crops, they respond well to rich soil, with a light, friable texture. Good drainage is a must, and if regular cultivation is given, it should be frequent and shallow. But a mulch is preferable, as with other vegetables and plants, to maintain an even soil temperature, retain moisture, and keep down weeds.

Dixie Butterpeas are rich in protein and carbohydrates. They are an excellent source of vitamins B1 and B2 and, particularly when green, a good source of vitamins A and C. The dried beans also contain some vitamin A.

Another unusual legume crop for the home gardener with a taste for something different in beans, is the broad bean, also known as the vicia fava bean, or just fava bean, Windsor bean, horse bean, and giant butter bean.

While Dixie Butterpeas are shaped like peas but taste like a richly-flavored bean, broad beans have a flavor reminiscent of peas although resembling large, dark, reddish-brown limas with a light stripe or eye. These shelled beans may also be cooked either fresh or dried, though they may be a bit coarse in the latter stage, developing a tough skin which requires longer cooking.

The standard variety is Long Pod, the historic original bean of England and Europe, which also is hardier than the other bean crops. It may be planted as early in the spring as the ground permits, since it does not like the summer heat. A heavy yielder of 7-inch pods, each containing 5 to 7 beans, it has the same general cultural requirements as other beans and other legumes.

Being a legume, the broad bean will, with the aid of its bacterial assistants, build up the soil in which it is grown. Is it wise to overlook such a valuable soil-enriching cover crop? It may also be said that every seed germinates, and the plants grow rapidly and robustly, brooking no opposition from such adverse factors as weeds.

Still another out-of-the-ordinary legume is the Asparagus Pea which also thrives best in the cool, moist, early growing season. It has very attractive reddish-brown flowers which, resembling sweet peas and other legumes, are followed by odd squared-off flared pods, best when picked young and tender and cooked whole.

Home gardeners should also raise a few shell beans, peanuts and other legume crops not too commonly grown on a small scale. Rich in food value, such home-grown vegetables have a delicious, different and distinctive flavor when compared to the store-bought kind. This is particularly true of peanuts, which may be raised as a home crop far north of the regions where they are grown commercially.

Though peanut pods may be planted whole, it's better in a home garden to shell them out first to prevent crowding and to encourage quicker germination. Don't cover the yellow blossoms with soil because you may reduce the yield. Early fall frosts won't hurt the peanuts after they have formed, but allowing the plants to be hit by frosts and turned brown will actually simplify harvesting. After they are thoroughly dry, when the pods are ready for use, the plants should be returned to the garden for the valuable organic matter they contain.

Ordinary snap beans will provide shell beans when they have fully matured. Some well-known bean varieties and types particularly well-suited for shell beans include navy, Great Northern, White Marrowfat, red kidney, and horticultural. Horticultural beans are especially good as green shell beans mixed with green snap beans, or with fresh sweet corn as succotash. The others are more frequently dried, though navy beans in particular also make excellent green shell beans. Dry shell beans should be heated in the oven for an hour at 140 degrees before storing in airtight insect-proof containers.

Seeds for all the bean varieties cited here may be obtained from most professional seedsmen; Dixie Butterpeas and Asparagus Peas are available from George W. Park Seed Company, Greenwood, South Carolina.

—Richard L. Hawk

Gain Time for Limas

In the North, necessary time may be gained by starting lima bean seeds in a hotbed, greenhouse, or even on an indoor window sill. Since a long taproot is involved, seeds are best started individually in 3- or 4-inch pots. The productivity and table value of pole varieties make this little extra effort pay handsomely.

Out in the garden, this is one vegetable not worth much gambling for time. Do not plant seeds until the soil is warm; two weeks after the last frost date is usually recommended. Then, planted an inch deep, seeds are sure to germinate, and early growth is rapid. Several times I have read that the soil for all types of beans should be somewhat poor. I had enough near-failures following that advice never to make the same mistake again. While beans should not be given the rich soil best for leafy crops, neither should they be starved. A good average vegetable garden soil gives excellent results.

King of the Garden limas can be trained to grow on a plastic-coated wire trelliswork.

Beans, like most vegetables, do require abundant water to be at their eating best. Certainly in these past years of drought, every good gardener has learned the value of a mulch. In the case of beans, wait until the soil is thoroughly warm before the mulch is put on, then apply it thickly.

Plastic-coated wire in a two-inch mesh is sold in rolls of various widths to use either as a fence or trellis. I have found this ideal for supporting all sorts of vegetables in all sorts of locations, including cucumbers, peas, tomatoes —and limas.

For the most dependability under a variety of conditions, 4 varieties of lima beans are basic. Fordhook 242 is the choice for a bush variety that matures large beans earlier than any pole variety. Sieva, the baby lima, is the earliest of the pole beans. King of the Garden is the old favorite large-seeded pole variety. I thought its taste couldn't be matched — until Prizetaker came along. This combines superb flavor with extra-large beans on vigorous, leafy vines.

—Richard D. Roe

Keeping Out Bean Rust

In recent years, according to USDA plant pathologists, bean rust has become a serious disease of pole beans, particularly in southern portions of the Midwest. Precautions recommended include not planting in soil where rust has infected a previous crop; choosing varieties which are rust-tolerant (White Kentucky Wonder 191, U.S. No. 3 Kentucky Wonder, and Dade); avoiding highly susceptible varieties (Blue Lake, McCaslan, Kentucky Wonder); and considering prevailing winds when choosing a planting site if there is a possibility of spores blowing in from an infected field. Since rust spores can be carried long distances, a long crop rotation is also advisable. The fungus lives over winter in old bean stems, but does not survive more than one year and is not seedborne.

BEET

Colorful Gem of the Garden

The beets I grow are one of the main crop varieties — the Detroit Dark Red. I must confess, however, that I selected it because I liked the name. Now I find that it really thrives here in the St. Louis area, with little attention needed except a good watering occasionally during exceptionally dry weather. Beets, though, do well over the entire United States—from spring to fall in the northern areas, spring, fall and winter in the extreme south.

The soil preferred by beets is a loose, friable, fertile one with a pH from 6.0 to 7.0. My garden soil is 6.0 with a 6 percent organic matter content. Nitrates test extremely high and so do the phosphates, calcium and magnesium, while iron is medium-high. In many cases the addition of garden lime is advisable, as beets do not do well in an acid soil. Another thing to avoid is use of raw manures, which adversely affect several root crops.

Some experts recommend that beets be worked frequently, but I do not find this necessary since my soil is loose and friable—and I mulch as soon as they reach a few inches in height. Furthermore, I've had no failures, even when my soil was much less friable.

I plant beets as early in the spring as the soil can be worked, covering the seed with about one-half inch of soil. Seeds are planted close together in rows 12 inches apart—somewhat closer than generally recommended. Planted in 12-inch rows, the beet tops soon meet between the rows, which shades the soil, helps to conserve moisture, and retards weed growth. Applying a mulch will give additional protection, while at the same time adding some fertility.

I begin harvesting as soon as the leaves on the young plants reach a size suitable for salads. Beet tops as well as the roots are valuable additions to cooked greens. My beet harvest is a continuous process from spring through the summer into late autumn. Actually, the harvest is at the same time a thinning process, with space

Beet harvest can be continuous from spring through summer into late autumn. The leafy nutritious tops are rated high in vitamins.

tually get as much as 600 percent germination from them, whereas 80 percent is generally pretty good for most vegetables. Once you've harvested the young rosebud-size beets and left space, the late sprouters are free to develop into big ones. Chard, though, should be thinned to a foot apart for best growth, but you can eat the succulent thinnings along the way.

Of course if you plant them in earth that is soggy and frosty or a hot dry clay, they may never germinate—but then neither would anything else except certain weeds. At 70 degrees F and with ample moisture, they will germinate in 3 or 4 days. As temperatures approach freezing, they will take several weeks.

left for the large-size beets to keep growing.

For big beets and heavy production, plant the Detroit Dark Red (59 days) in fertile soil; thin by harvesting from 8 to 12 inches apart, and let them grow until danger of freezing. When the temperature is due to drop below 30 degrees F, I cover my beets, using straw, plastic or rags for protection and removing the covering when the freezing danger passes. Cardboard or paper may also be used.

—Elmer L. Onstott

Beets and Chard—Two for the Vegetable Row

So easy to grow is this pair of tasty vegetables that sometimes more plants come up than you planted! How come? Well, the corky seed of both beets and Swiss chard is really a compound unit of anywhere from two to 6 seeds. And although not all germinate at once, you could even-

Then there's another advantage: these two vegetables are among the most colorful. In case you want to slip them into the flower garden without their looking utilitarian, they not only go in well, they look great doing it. Their red and green leaves make handsome foliage plants, especially the new Burgundy rhubarb-red Swiss chard with stalks as bright as cherries. A few varieties of both beets and chard have foliage that is only green. Different varieties make pleasing patterns. They don't have to be planted in rows. Clumps grow just as well if the soil is rich and humusy. Figure at least an 18-inch clump for each person who will be eating beets, and 3 or 4 plants for each chard eater.

The difference between the two plants is that the beet develops a ball-like root (technically an underground swollen stem), but has rather small leaves and skinny stalks. Chard has fibrous roots that aerate the soil—but no edible ball—and its leaves are large, its stalks wide. Both wait until the second spring to shoot up enormous flower spikes completing their biennial cycle.

Most of us don't let beets grow that long unless we want to use them for... d. By then the round red ball has lost its sweet succulence and would still be woody, even after hours of cooking. The best flavored are young, golf- or tennis-ball size, although some varieties are good until they are as big as baseballs, especially if grown in rich humusy soil.

The same planting instructions apply to both beets and chard. They are considered cool-weather vegetables, doing best from fall through spring where ice never forms on the birdbath, spring through fall where it does. Beets are somewhat heat-tolerant, however, and deep-rooted chard is quite heat-tolerant compared with lettuce or peas, especially if your soil has plenty of humus. Beets are paler red and not as sweet if grown through hot summers. Chard gains amazing vigor in fall's cool temperature.

A soil that grows daffodils or gladiolus will grow beets. Plenty of humus and, unless they're grown during a very rainy season, a mulch on top of the soil contributes to rapid growth and quality beets. In clayey soil their root balls rise part way out of the earth, but this doesn't harm their flavor appreciably. Unlike carrots, they don't become deformed in clay unless it is very heavy and compacted or full of stones. Fairly neutral soil is best. Good drainage is essential as with all root vegetables. Manure and bone meal dug into the soil several weeks before planting is the traditional and most reliable method of fertilizing. A top-dressing of wood ashes after beets are up adds needed potassium and helps control slugs and snails where these are a problem.

Chard needs the same kind of soil. The deeper its humus content and root penetration, the bigger the leaves. But the top-dressing should be rich in nitrogen, such as cottonseed meal, dried blood or soybean meal. Since chard is a long-season crop, manure can be added several times during the life of the plant.

Both of these vegetables like lots of deep water. Generally, water them longer than you do the lawn, but half as often.

You can pick chard leaves all year if you take outer ones only and let the center continue to grow. If chard winters satisfactorily in your area, you can begin picking again in early spring before the top growth begins. After blooming, it's finished.

Beet leaves can be harvested gradually, too, and you may have your beets and eat the leaves as well, but they are smaller than chard. If you have space for both, plant both because the difference is worth it. If you have space for one, plant beets and use the leaves, too.

Chard's big leaves give quantity "spinach." Even though they feel leathery compared with the lettuce-thin other vegetable, they aren't tough when cooked unless they are old. They don't tolerate days of lying limp in the supermarket nearly as well as spinach does. That's why you seldom can buy it—and home-grown is best-grown.

Swiss chard also makes good imitation celery. The stalks—often an inch wide—can be substituted in almost any cooked, but not raw, celery recipe. Or they can be left whole, cooked like asparagus, and covered with any sauce used on asparagus.

—Marguerite E. Buttner

BROCCOLI

You Can Beat the Weeds!

I always had trouble with weeds growing against my fence, especially one called Creeping Charlie which was so hard to remove because as it grew it became entangled in and under the mesh. The matted leaves blocked out sunlight, and the weeds just did not grow in that area. Later on, as I set out cabbage and broccoli plants, I took forkfuls of matted leaves from the top of the pile and used them as mulch between the rows of plants. The sun baked the top layer to a hard, solid mass, but underneath the earth stayed moist. (I checked, by occasionally turning over a small section.) That summer I had *absolutely no weeding problems!*

My broccoli plants became the envy of our neighbors. About 3 feet high with dark-green foliage, they yielded huge heads of broccoli that tasted wonderful. And since I abhor sprays and just will not use them, what was astounding to my neighbors was that their broccoli was covered by caterpillars even though they used spray liberally, while I did not have one caterpillar on mine! I put 45 boxes of broccoli in the freezer besides the innumerable times we enjoyed it as a fresh vegetable for lunch or supper. Side shoots grew after the center bunches had been picked, and I was still going out to pick them practically up till Christmas. I can remember picking fresh broccoli with my winter coat on and the first snowflakes falling gently around me.

—Joan Pierson

From Cultivation to Cooking

Because broccoli needs a continuous water supply we mulch our plants with hay so we won't have to do any watering during drought spells that have prevailed the last few summers. Although not greedy, it does return higher yields when the nitrogen content of the soil is high. So after the first heads are cut I like to side-dress with well-rotted manure, which encourages larger yields through the entire harvesting season.

The broccoli plant grows with a thick main stalk eventually topped by a central cluster of tiny, green flower buds. Harvest these while the florets are still tight buds, otherwise they open into flowers, and most of the delicate flavor is lost. The stems and leaves are edible too, even if not quite as tender as the sprouting tops. After the first cutting, any side shoots that appear should be kept cut so that you have a continuous crop for many weeks.

Because it is much more perishable than head cabbage or cauliflower, broccoli has superior flavor for eating if you cut it and cook it immediately. This is one argument for growing your own, because the supply at the supermarket is probably travel-worn.

My favorite method of cooking broccoli is to stand it up in the bottom of my double boiler with the stems immersed in water. The flowerets steam tender under the inverted double-boiler top. I save the cooking water and use it in a sauce or soup.

To prepare for cooking, trim coarse stalks, and peel up to the florets. If you are serving it whole, split thick stalks to hasten cooking time. Rinse and drain, then tie into bundles. Whole broccoli cooks in 15 to 18 minutes. Overcooking dulls both the color and taste. For less fancy serving and quicker eating, it may also be sliced and cooked quickly in salted water. I usually reserve the florets and add them for the last few minutes of cooking so they won't overcook.

Try it also the Italian way. Boil it crispy tender, drain and saute gently in hot olive oil until it is light brown. Serve sprinkled with grated Parmesan cheese. If you prefer your vegetables raw, chop it fine combined with tomatoes, and serve with your favorite dressing or sour cream.

—Joan Lindeman

BRUSSELS SPROUT

A Crop for Cool Weather

Brussels sprouts are considered by many people to be a rather chancy crop for the home gardener. I find nothing could be less true if the particular requirements of the plants are met. Actually, Brussels sprouts are a hardy, slow-growing member of the cabbage family and are primarily a fall and winter vegetable. They must have cool weather to mature in; overlooking this fact is the basic reason for most failures.

Sprouts will not mature properly in sections of the country that don't have frost occurring; but any section that can grow late cabbage can enjoy them. The main thing to remember in the cultivation of this delightful vegetable is that it does its best growing in cool weather. The first year I tried to grow it I started my plants too early in the spring and had them maturing during warm weather. The result was quite discouraging. The sprouts were not tight, neat, miniature cabbage heads but loose-leaved, tough, green knobs of nothing.

I start my Brussels sprouts in late May, sowing them thinly in a shallow drill and covering them lightly. After they have their true leaves I thin them to about 3 inches apart. When they are about 3 to 5 inches tall I thin them again to stand about 3 inches apart. The thinnings I transplant along the rows until I have utilized all of them, leaving at least two feet between rows. They grow fairly slowly during the heat of summer, but as the cool weather approaches they finish their growing with a rush.

26

My sprouts are grown in a mulched garden well enriched with both cow and chicken manure, as I have learned that lack of sufficient moisture and nourishment will make for a poor yield.

To harvest sprouts, start at the bottom of the plant as these buds mature first. I pick the lower leaves off beginning in late September, as this seems to concentrate the plants' energy on the immature buds. Don't delay in picking the sprouts as quickly as they mature or they will toughen and lose their subtle flavor. What a loss this would be, as this flavor is the most delicate of a clan not necessarily noted for delicacy! Being a cool-weather crop they taste best when nipped by a frost or two, but won't survive more than one good hard freeze. If they should freeze, they are still edible if allowed to thaw out slowly.

—Joan Lindeman

Stretch the Vegetable Season

Because Brussels sprouts take patience to grow—up to 100 days to mature—I start mine indoors on the windowsill and transfer them later to the cold frame. I set them out in our New Hampshire garden in early June. (Further south, where frost-free weather arrives earlier, these steps can be taken sooner.)

Like all leafy plants, these cabbage-family members like plenty of nitrogen. I stirred a generous supply of cow manure into the soil before putting in the plants. Compost, dried blood, cottonseed or soybean meal fertilizer can also be used. And I gave them plenty of water, too.

Once the plants catch on, they grow woody, broomstick-like stems that reach a 2- to 3-inch diameter.

Leaf stems shoot out in tiers like the spokes of a wheel and support heavy, gray-green foliage.

At the base of the leaf, where it joins the main stem, small sprouts or "buttons" appear. To encourage these sprouts to grow and mature to edible size, the leaves should be snapped off from the main stem. The best time to remove them is when the sprouts have formed into a tiny rosette. Removing the stem and its heavy foliage lets more energy and sunshine go toward developing the sprouts — which are actually the buds of the plant.

Taking off the lower leaves was a weekly chore for me until the plants reached about 18 inches tall. I did not remove the top leaves, but allowed them to develop so that plants would not be destroyed. It was interesting to watch the tiny rosettes on the lower part of each stem form into what looked like small cabbages.

In early July I gave my plants a booster feeding of cow manure and watered it in well. A mulch of grass clippings or hay helps keep down the weeds and conserves moisture. Usually, almost all of my mulching material is being applied to the squash and pumpkins at that time.

By late August I was able to pick some edible sprouts, and by late September the crop was abundant. As the frosts came, the sprouts improved in flavor. They are best picked at 1½ to 2 inches in diameter.

If you want to continue your crop into early winter, transplant to a deep cold frame or to a large bucket that can be placed in a cool cellar or garage.

—Walter Masson

CABBAGE

Shortcut to a Double Harvest (or—Two Heads Are Better Than One!)

Two heads of cabbage from one plant—that's what we get! A good second crop needs good soil. It's worth trying.

We planted radish and early head lettuce in alternate rows 18 inches apart, as early in our Colorado spring as possible. After they were well up, we planted Golden Acre cabbage plants 30 inches apart in the radish rows. About a month later when it warmed up and all the radishes had been pulled, we planted a pepper plant between each cabbage. Lettuce should all be gone before cabbage and peppers need the room.

Wood ashes dusted several times during the summer kept the white butterflies of cabbage worms away; fresh ashes also make these worms shrivel up and die. Gets most aphids, too. Mulched with planer shavings in July. A nearby sawmill gives us shavings and ashes for the hauling. We get a 25-gallon drum of ashes at a trip, plenty for our needs.

We began cutting cabbage for salad early in July and had plenty yielding all the rest of the year. After cutting a cabbage head, many small heads grow on the stalk. None can get big. We come back later the same day and cut the stalks off lower down, trimming to just 3 leaves. After small heads start, leave only one to a plant and it will usually make a marketable head, only a little smaller than the first head was. These second heads are long, not as round as Golden Acre usually is. By getting two heads from each plant we do not have to seed late cabbage.

Maybe the interplanted peppers help keep bugs off the cabbage. At least they do not interfere with each other, and we get a larger crop than on the same ground space planted separately.

—Charles W. Lindsey

Time-Space Plantings

Since cabbage plants are so easy to start from seed, we grow our own in the basement each spring. We sow early and midseason varieties in shallow drills, using small flats of good garden soil. (Add no fertilizer to the seeding soil.) Cover seeds with 1/8th inch of soil, then firm down well. Next, the flats are watered from the bottom, and slipped into large plastic bags so that temperature and humidity will remain constant for good germination. Cabbage seed sprouts quickly at temperatures around 60 degrees. After all have sprouted the flats are removed from the bags and placed in a cool, south basement window. These seeds are sown 7 weeks before we expect our last hard freeze in May.

As soon as the seedlings develop their first true leaves, I transplant them into the cold frame where day temperatures reach 65 degrees F, with a night drop to about 50. Seedlings are shaded for two days until they catch on, then are exposed to direct sunlight. *Remember, cabbage plants*

28

need all the sun they can get, yet growing temperatures must remain on the cool side. This combination results in short, stocky plants that produce firm, massive heads. Watch the weather. When the cold-frame thermometer reaches 65 degrees, raise the glass slightly. Close it at sundown to hold soil heat so that night temperatures will not drop too low. Keep soil moderately moist at all times, but do not feed plants while in the cold frame, or they will grow too tall and lanky for good production. The stronger the transplants you set out, the better the crop of cabbage you'll harvest.

If you prefer to wait, seeds of midseason varieties can be started directly outdoors when you plant carrots. They will take mild frost with no ill effects. Late varieties are started outdoors July 15, and transplanted by August 15. In southern areas, they may be started earlier.

Seven to 8 weeks after seeds sprout, they're ready for transplanting to the open garden. Good quality heads depend on constant, rapid growth—and that calls for plenty of food and soil moisture.

Early varieties that form small heads are set 12 to 14 inches apart in rows 24 inches apart. Space larger kinds 18 inches apart in 36-inch rows. Use a dibble or trowel when transplanting. Wrap a strip of newspaper around the stem of each plant, then set it so that its lowest leaves are just a fraction of an inch above soil level.

I've found soil moisture is the most vital aspect in the production of sound heads. Don't allow the soil to dry out around roots; if it does, growth will slow down. Then, when rains do finally come, plants spurt into such rapid growth the heads often split. Conserve moisture and help keep soil cool by mulching the rows heavily with spoiled hay or other material as soon as heads begin to form. If irrigation is necessary, use a sprinkler until soil is soaked down to 6 inches. For exceptionally large cabbage heads, scatter a liberal side-dressing of aged poultry manure in the aisles as the heads start forming or before you lay down the mulch.

Early and midseason varieties may be harvested as soon as heads have gained sufficient size. Late varieties store well. Allow them to remain in the garden until a hard freeze is expected, then cut them off and store in a cool outdoor building. They may also be pulled up by the roots and heeled-in upside down in a well-drained trench in the garden. Cover with a thick layer of straw to prevent freezing. If you have no outdoor building, wrap them individually in wax paper, and store on shelves in the coolest part of your basement or attic.

One final hint: If you've hesitated growing and serving vitamin-rich cabbage because of strong cooking odors —don't. Actually, when cabbage is cooked properly there is no offensive scent of any sort. Strong odors occur only when cabbage is cooked at too high a temperature or in too much water.

—Betty Brinhart

Chinese Style

This quaint "cabbage," developed in the Orient, is really a member of the mustard family. Introduced to American gardeners by seedsmen about 1885, Chinese cabbage is not too commonly grown in the vegetable garden, perhaps because it is a cold-weather plant. And although it resembles Swiss chard with crinkly outer leaves, it heads up inside to form a tight leaf pack about 15 inches long that can be cut up and served boiled or in cole slaw.

To get a head start, I sowed seeds in April in composition pots in the cold frame, then transferred them to my New Hampshire garden in late May. My object was to get the plants to grow and head up as quickly as possible—before the heat of the summer set in. They have a shallow root system for the size plant that must be supported. Because of this, I gave them plenty of dehydrated cow manure as well as water.

As the plants grew, I thinned them out to 15 inches apart, with rows spaced about 2½ feet apart. By mid-July I had my first harvest of 18- to 20-inch-tall plants with firm center heads. Some that my family and friends could not consume fast enough went to seed. I eventually pulled these up to add to the compost pile, where they made beautiful bushy additions. Hot weather slows growth and develops seed heads.

Chinese cabbage takes 70-75 days to mature, and for this reason it should be started in the cold frame or directly in the soil of the garden as soon as it can be worked. If you live in an area where frost doesn't arrive until mid-October, you can try a fall crop by planting the seeds in August.

Organic food high in nitrogen (soybean or cottonseed meal, fish meal, etc.) and water are important to encourage the plants to grow quickly. Keep the area free of weeds. I use a mulch of grass clipping or hay to hold moisture in the soil and help keep down the weeds.

You can store harvested plants for several weeks in a cold place by cutting off the roots and letting the outer leaves remain on the plants. Some varieties can be quick-frozen.

My family prefers to eat Chinese cabbage boiled. Remove outer leaves and cut the inner head into one-inch slices, then boil as you do regular cabbage. Drain and serve seasoned and buttered.

—Walter Masson

Outwitting the Cabbage Worm

I tried spooning a little sour milk into the center of each cabbage. That for me was the perfect solution to outwitting the cabbage worm. Skim milk soured with vinegar is cheap, quick to make, and easy to apply; its effect endures for the greater part of a week; it doesn't drive away friendly bugs or birds, and it keeps butterflies off as well as do the sharper vegetable sprays. It doesn't get rid of every worm, but with my ever-increasing army of insect pensioners I wouldn't want a method of control that did. To go down the rows dropping a teaspoonful of sour milk onto each cabbage is just 5 minutes' work. It needn't be done the first thing in the morning when the dew is on them, but whenever there is a little free time.

Another very simple device I discovered by accident is a pool of water near insect-ridden plants. I leave the hose there an hour or so; birds, yellow jackets and mantids gather for the water, notice the insect food and go to work.

The thing that my research on cabbage worms taught me is that it is unnecessary either to surrender a crop to them, or to save (!) it by poisoning it. A solution other than the one I fastened on might be a better one for you. Decide for yourself. Cabbages are too good to be destroyed either by worms or by poison.

—Dorothy Schroeder

Pepper Spray

Mrs. E. C. Shaver of Houston, Texas, makes a safe insect spray from plants. "My vegetable garden is in the second year of organic gardening, so I still have lots of insects. My worst trouble is with cabbage worms. I have tried all the simple remedies and nothing gets rid of them. I was determined not to use poison sprays, so I looked around for something organic to use. I had several plants of a hot pepper that is so hot we can't eat it. I ground up several pods of the hot pepper, added an equal amount of water and a half spoon of soap powder to make it cling to cabbage and collards. I sprayed with the hot pepper water and my cabbage worms are all gone.

"I also tried the pepper spray on several insects of different kinds with very good results. It was very effective on ant beds, spiders, caterpillars and tomato worms. I was so pleased with the results of my pepper spray that I canned several jars of ground pepper to save for next spring when I won't have fresh pepper in my garden."

Covers for Cabbage

Worm-free broccoli and cabbage can be grown in your garden without spraying or dusting. One year we used protective canopies of black polyethylene netting. Because it kept the egg-laying parent moths away from the tender young plants, we had a perfect crop.

The netting, either black or clear, comes in rolls 7 feet wide by 100 feet long with one inch or 7/16-inch openings which are almost diamond-shaped, like chicken wire. We found the 7/16-inch size kept the white moths completely off the plants, thwarting their every effort to lay their eggs where they would hatch into crop-destroying worms.

The double row of cabbages is protected by a cylinder of hoops and mesh against worms.

At first I thought the 7-foot width might be awkward, but it was more than satisfactory for the cabbages and broccoli. Lightweight and flexible, it handles and drapes easily and may be cut with ordinary shears—a big advantage over chicken wire. With reasonable care it can be used repeatedly, for several seasons.

I planted the broccoli and cabbage in double rows because I thought the 7-foot width would give me room to spare. I later found this was just not the case because I did not allow enough planting height under the tent. The moral of the story is plain— be sure to leave plenty of room for your plants to grow!

I allowed 24 inches between rows and between the plants. To support the netting, I cut 8½-foot lengths of #9 steel wire commonly used for fence bracing and bent each wire into a half-moon. The ends were pressed 6 inches in the soil until they felt secure. (This depth can vary with different types of soil.) Spaced 6 feet apart in the row, the hoops looked like giant croquet wickets 4 feet wide and 26 inches high, which I later found out was not nearly high enough.

I tied binder twine from the top of one hoop to the next to support the netting without sagging, although 16 feet of it weigh only one pound. Because the moths are persistent in their search for a spot to lay eggs, I was forced to secure the bottom edges of the netting, making a perfect seal with the ground.

On one side of the row I scratched small furrows with my hoe and buried the edge of the netting. On the other side I used grocery-store string to tie the mesh to the bottom of the wicket so I could remove it to permit weeding. While it was easy to untie the string, flip up the netting and remove the weeds, I plan next season to mulch heavily—which should eliminate this chore.

Of all the methods we have used to achieve worm-free broccoli and cabbage, the polyethylene plastic netting is much the best. It's foolproof and the cost is quite low. Most important, it reduces insect infestation in the rows practically to zero.

—Charles F. Jenkins

Blender Concoction for Cabbage Pests

This is the first year we haven't had our cabbage-family crops ruined by the white cabbage moth. A friend advised us to put salt in each hill of soil when the plants are set out. We did this, plus watering each plant with a dilute solution of our special mixture whenever I noticed the white moths hovering over the garden. It has also been very helpful when the squash and cucumber plants were almost ruined by the spotted cucumber beetle. After my "treatment," the beetles moved elsewhere.

I use the blender when whipping up my concoction—and even the dogs leave the kitchen! I cannot give specific amounts of each ingredient;

let each individual vary his own.

Ingredients used: fresh spearmint, green onion tops, garlic, horseradish root and leaves, hot red peppers (dried, were all we had on hand in June), peppercorns and water.

Blend all this in the mixer, then pour into a gallon jug. Add about one cup of inexpensive liquid detergent. Make as many batches as you like. When you want to use this mixture on your plants, use about ½ cup insect repellent mixture to about one quart of water. I have an old dipper that I use to give each plant a good dose.

Up to now, the benefits have been many. I never like to use poison sprays, as ours is strictly an organic garden.

—Frieda A. Starcevich

Tin Can Gardening

Tin cans can be an important addition to your cabbage patch. I use # 10 gallon-size food cans and cut out both tops and bottoms. When I sink them into the soil and fill with water, I have small but effective reservoirs!

I set several hundred cans in rows the length of my garden, planting in the rows between them. I stretch a line to keep them straight, set them in about 3 feet apart, leaving the top rim about one inch above ground level to keep the soil out, and stuff a wad of marsh hay, straw or lawn clippings into the can. I spade this material under the following spring when it is well-composted, using a trowel to dig out any soil that has fallen in, and filling the can with a wad of new hay.

I have been able to improve on this system by digging a little larger hole, filling it partially with compost, setting in the tin cylinder, and then filling up with good garden soil. I find I am getting much better results be-

I work it something like this:

For early and cool-weather crops like head lettuce, cabbage, cauliflower and broccoli, I set out one plant at each irrigation hole as early in the spring as possible. Then, when it is time to plant warm weather things, I set them in on the opposite side of the same tin can.

In theory I can always get at least two crops. But in practice I get 3, and sometimes even 4 crops, all at the same watering hole. For instance, I plant head lettuce and early cabbage in the same row on opposite sides. When the head lettuce is gone, I plant peppers, gumbos, tomatoes, melons, or cukes; and when the early cabbage is gone, I plant late cabbage or broccoli, or even early sweet corn, planted late. There is no end to combinations that can be made of early and late crops in this companion cropping, all utilizing the same tin-can watering hole for irrigation.

—John S. Park

In a test involving electrocultural principles, we planted celery between rows of tin cans at the Organic Experimental Farm, also clery without cans. Celery with cans grew 34 percent heavier than the control group.

cause the hole gives off liquid fertilizer every time I fill the can with water, enriching the surrounding area. Planting in hills on both sides of the sunken cans, I start early and late varieties opposite each other, companion style, for continuous cropping.

Editor's Note—In addition to irrigating his garden efficiently and thriftily by filling the cans so each acts as a small reservoir, Mr. Park is also attracting atmospheric electricity to the growing row.

CARROT

Fresh-Dug at Snow-Melting Time

Leftover snow lay in sheltered corners of the yard when my husband spaded the garden last spring and found two carrots—fresh and crisp and sweet—that had been overlooked the previous fall. When you consider that the ground had been frozen to a depth of 3 feet and that our northern Minnesota wilderness temperature had reached 45 degrees below zero, the healthy roots took on the appearance of a minor miracle.

Our next harvest of Improved Long Orange was so bountiful that, last October, after we had exhausted our supply of sand and containers suitable for winter carrot storage, two rows were still not dug. We decided to try to hold them over in the ground. After all, if two carrots could survive, why not two rows?

We clipped the tops about a half-inch above the roots and covered the rows with a wide-, 3-inch layer of coarse sawdust, which we hoped would be a protection against extreme cold snaps that might come before the snow. The glittering, white blanket was 3 feet deep by December first and did not begin to melt until the latter part of April. As soon as the garden area was bare, we scraped away some of the sawdust. Bright and green, young carrot leaves peeped out.

The thought of tender carrot strips was mouth-watering because our stored supply was long gone and it does not pay to have them mailed to us from the village in winter as they stand a good chance of being hopelessly frozen on their 45-mile trip. But the ground was still ice-hard; not even a pick would break it up! Regretfully, we replaced the sawdust as a deterrent to strong sprouting which might result in seed-stalk formation and inedible roots.

In mid-May, the ground was fit to dig for this year's carrots and, on the day we planted, we ate the first of those held over from last year! On June 15, when leaves were beginning to poke up through the hindering sawdust layer, we ate the last of the roots, which were still perfectly firm, and replanted the two rows.

Our sand-stored carrots, which are watered regularly and kept at an above-freezing temperature, put out many tiny rootlets and show considerable sprouting. By March, their texture is less delicate and their flavor less sweet, but the moist storage prevents their wilting. Apparently the dryness and very cold storage temperature of those left in the ground held back growth so that there was no deterioration.

Why don't you experiment with a few of your own carrots, if your garden is in an area where the ground freezes to at least a depth of a few inches? If your snow cover is not dependable, a thick layer of any close mulch which extends somewhat beyond the rows should prevent either too hard freezing or frost heaving. If this works for you, you will find that the sight of those bright-orange roots, fresh from the bedraggled earth of early spring, is the sort of thrill no vegetable gardener should miss.

—Helen Hoover

Give 'Em a "Coffee Break"

Tired of brown worm trails spoiling your carrots? Foil that pesky fly whose larvae ruin them. Give the carrots a coffee break!

Here's an easy, safe way to keep this pest at bay. Mix your package of carrot seed with one cup of fresh *unused* coffee grounds. Plant the coffee with your seeds. It percolates enough coffee odor during the growing season to fool the nosiest of carrot flies. It won't flavor the carrots as naphthalene and creosote do, and it's nonpoisonous.

If you prefer to broadcast your carrot seed, mix a couple of quarts of fine peat moss along with the coffee. Simply spread the seed-coffee-peatmoss combination over the top of the prepared area and tramp it down. Dampen and it's ready to grow.

—Ruth W. Godsey

The carrot fly can be foiled by the aroma of coffee grounds mixed with carrot seeds.

Molasses Mix Beats the Worm

Here's my method for keeping worms out of carrots.

When the new carrots are two to three inches high, you take one quart "stock" molasses (crude black molasses) to 3 gallons of water and pour over the rows of carrots. Repeat if necessary. I do this only once, never have any worms, and I have even had my carrots in the same row for 3 years.

—Mrs. A. P. Lachapelli

Well-Formed Carrot Roots

The soil in which carrots are to grow must be well prepared and enriched. It should be deep, mellow and friable. Although the carrot gets along well in almost any type of soil, this should be free from lumps and stones which often force the roots into deformities and cause them to split. Nematodes are also a possible frequent cause of misshapen carrots. Large applications of humus may prevent nematode damage.

A good supply of humus from a well-made compost heap will do much to put the soil into condition. Exhibition specimens are sometimes grown by drilling holes in the ground and filling them with a mixture of equal parts of compost, leaf mold and sand. Apply finely ground limestone when soil tests show it is needed to correct excess acidity.

Deep cultivation frequently injures carrots, since their feeder roots remain near the surface. Use a fine-shredded mulch to control weeds. Thin carefully when carrots are about half an inch in diameter so that remaining plants stand approximately two inches apart.

CARROT

Sow Carrots Thin

Some gardeners prefer to sow their carrot seeds thinly to eliminate thinning later. This can be accomplished by mixing the seed with a tablespoon or two of fine sand before planting. Although this does save work, I don't like the idea. As a rule, not all carrots sprout at the same time. Those that sprout first produce the largest roots. I like to sow my seed thickly, then weed out the slow, weak ones. By doing so, almost every carrot I harvest is a large, straight one.

One of the most important steps in growing nice carrots is proper thinning. Carrots will tolerate some crowding, but do best when given plenty of room. I begin thinning mine when they are 1½ to 2 inches tall. The stouter plants are allowed to remain, while the smaller, thinner ones are removed. This usually leaves one inch between plants.

A month later, I go back and thin the rows again so that the remaining plants stand 3 to 4 inches apart. The roots of those that I have removed are of reasonable size by then and go into my salads or into hot dishes and soups. In order not to disturb the roots of those remaining, I like to pull out the undesired plants right after a shower when the earth is good and damp. Carrots that are properly thinned will produce as much as 125 pounds per 100-foot row—far more than unthinned rows ever could.

After the last thinning, I loosen the soil in the aisles as deeply as possible with a grub hoe. These aisles are then mulched heavily with dried grass clippings or old hay. I take special pains in pushing the mulch right up against the plants to shield the carrots from the sun as they grow larger. Soil can also be used for this purpose.

Exposed carrots turn green, and take on a bitter, unpleasant flavor.

They definitely need a steady supply of water. I have made it a practice to apply at least an inch of water once a week if rains are few and far between. In extremely hot and dry weather, I water the plot twice a week. If you use mulch, check under it often. As soon as the soil feels the least bit dry, water.

—Betty Brinhart

Some Down-to-Earth Advice

When you're ready to fertilize your carrots, avoid raw manure at planting time. It causes forked roots and rough texture. Raw manure may be put on the fall before and dug in, which is what I do over the whole garden area. For many crops, including carrots, that's enough richness. If you haven't put on raw manure previously, it is safe to mix dried manure with the soil when sowing carrot seed. But better, I think, is compost, well-decayed compost which has fertilizing as well as soil-lightening qualities. A small amount of fine humus goes a long way if placed in the bottom of the drill or sprinkled then firmed over the seeds. Wood ashes are beneficial for carrots, too. Mix thoroughly into the soil before planting, or spread along the row on each side. Ashes also help guard against wireworms.

The books say, "Plant carrots as soon as the ground can be worked." That's a little misleading, for the same is said of peas, the first sowing of which I usually make here in the Boston area at least 3 weeks before that of carrots. Soil should be dried out more for carrots, should be crumbly, not compacted or "clingy." A rough furrow is enough for peas, but soil must really be worked for carrots.

Earlier plantings may be made in light, sandy soil. My first sowing is in late April, then small succession sowings until a final big one in mid-summer.

Sow carrots in shallow drills, thinly in spring when germination is good, thickly in hot, dry weather. In summer, watering may be needed until seeds sprout. If drill is covered with a little loose hay (not so deep as a mulch), watering through it will keep fine seeds from washing away. It also keeps the soil surface from crusting so that sprouting seeds can't break through.

Space between rows should be mulched more deeply. Thin seedlings when one to two inches high, then again in a week or so. Plants meant to attain full size should stand two to three inches apart in the row. Carrots can be used at any stage, and "fingerlings" are most delicious.

Peas have soil-loosening qualities which benefit carrots. Plant as companions (in neighboring rows) or plant carrots after peas. My last sowing of carrots is made about July 20 in rows recently vacated by the last peas.

Carrots can be used at any stage of growth, but the "fingerlings" are the most delicious.

Wherever you live, the date of this important last sowing depends: (1) on the average date of the first frost (carrots won't grow much if any after that); and (2) on the number of days required for good size or maturity of a particular variety. (Some may take almost 3 months; all take over two.)

I sow more carrots at this time than at any other, yet later on always wish I'd sown even more. Why? Well, such carrots are intended for late fall and winter use when few other vegetables can be harvested in cold climates, especially vegetables so rich in flavor and nutrients, and so easily kept—right in the ground. Carrots are semi-hardy, will rot if not mulched with leaves or hay over the rows. But delay covering until after several hard frosts, and mark rows with tall stakes to be seen above deep snow. I have dug up carrots in 20-degree weather when unmulched ground was frozen hard—even under snow. It's practical, though, to dig up a week's supply at a time. I've found carrots kept in the cellar for long periods, even if covered partially with dirt, are far less crisp, juicy and sweet than those dug from the row and used at once.

Carrots should be mulched lightly in the row, but more deeply in the middle areas.

CARROT

Carrots are semi-hardy, and will rot if not covered with hay or leaves in the late fall.

This hardy vegetable also cans and freezes very well, much better than do peas, meaning they really taste fresh. Nevertheless, I prefer to get carrots right from the row as many months in the year as possible. I've harvested the last of a previous year's crop in late March, then pulled the first spring "fingerlings" about June 1! Further south, a gardener can do better.

—Ruth Tirrell

Sweetest Carrots Ever

For years carrots grew as one of the staples in our garden, which we enjoyed immensely during late fall and winter when many fresh vegetables are no longer available. Each year we planted in the usual way, sowing the tiny seeds as thinly as possible. Even with careful planting, however, we still had to thin the little seedlings to give those remaining enough growing room. Usually we did this when the seedlings were so small that they had not yet attained their familiar orange color, but resembled instead a host of upthrust white needles.

One year, due to a busy season, we found ourselves in the carrot row much later than usual. Many of these young carrots were already ¼ to ½ inch across and orange-hued throughout. It seemed a waste to throw them on the compost heap, so I brought my "finger" carrots into the house and *tried* to peel them. This effort met with no success. So, with misgivings, I cooked them skin and all. When cooked, I added seasonings—and to our surprise, we enjoyed the most delicious carrots we'd ever eaten. In fact, finger carrots, I believe, have such a delightfully different taste from regular carrots that they seem a completely new vegetable. The skin is so tender that it seems nonexistent and, of course, more of their vitamin A content is retained by leaving it on without peeling.

Of many good methods of winter storage, we prefer keeping our carrots in large stone crocks of sand from which they can be removed as late as the following April, still firm and bright-colored. Even at this late date, the taste is far superior to that of any so-called fresh carrots to be found in the supermarket.

—Inez Grant

CAULIFLOWER

Trouble? It's All in Your Head!

Blanching is the secret to tender, snowy-white heads. Those heads that are not protected from the sun and wind as they mature, develop a brownish tint, and undesirable flavor which hampers their quality.

When the tiny white balls, or curds, are just forming, they are well protected by a tight whorl of small, crisp leaves which block out the light. But, as the head grows larger, it becomes exposed. As soon as it can be seen through the leaves, gather up all of the surrounding leaves and tie them securely over it. This will protect the head so that it will continue to grow rapidly without injury to its color or flavor.

If you have several plants, you might tell the difference between the time they have been tied up by the color of string or cloth used. That way you can tell which should be matured without having to untie them all. As a rule, heads will mature from five days to two weeks after being tied up, depending upon the weather. They will mature the fastest on hot days, so keep an eye on them by checking every two or three days.

Many excellent heads of cauliflower have gone to ruin in the home garden because they have not been harvested at the peak of their goodness. Watch the heads carefully, for they mature faster than you think. When ripe, the head is compact, very firm, snowy-white, and even across the top. It is far better to take a head a little earlier than later, for an overdeveloped head has a strong flavor. Some gardeners make the mistake of delaying harvest because they think the head will grow larger if left alone. As soon as a head reaches maturity, it does no more growing. It will only age if not taken.

When harvesting, cut the head off at the very base if you intend to use it right away. But, if you want to keep it for several days, allow a whorl of leaves to remain around the head to keep it fresh longer. These heads may be wrapped in waxed paper and kept in a refrigerator until used, or placed in a cool, damp basement on the floor.

—Elizabeth Matvey

CELERY

At Last—We Grew It!

Five years of failure trying to grow good celery, but we finally learned the following method: The one most helpful instruction was that the celery plants must never be allowed to dry out, never. Our experience showed that this is probably the one most important requirement in growing tender, not stringy or hollow celery.

1. Green Light, original strain, was chosen because the catalog recommended it for home gardens. It is erect and compact and not too tall, thus making the stalks easy to handle when harvested.

2. The seeds were planted indoors on March 15.

3. The seedlings were transplanted to a cold frame early in May. They were set about 4 inches apart in rows 4 inches apart.

4. The plants were again transplanted in late June, this time to a specially prepared bed which was part of a flower border near the kitchen where water is readily available. The soil in this bed is rich topsoil. A trench approximately two feet wide and 25 feet long was dug to a depth of 4 or 5 inches. A double row of young celery plants was planted in this shallow trench and watered thoroughly. The plants were set about 15 inches apart in the row. The tiny weeds in the trench were removed with a hoe a couple of times during the next 3 weeks, while we waited for a good soaking rain before mulching. The topsoil piled at the sides of the trench was gradually brought down around the celery plants. After the second hoeing, the celery bed was level with the surrounding area. After the rain finally came, the entire trench and about a foot around it was mulched heavily with field cuttings. During the entire growing period, indoors and outdoors, the plants were supplied with water regularly so that the soil never dried out.

5. In September and October we harvested the huge stalks of celery. Some of the stems were ⅓ to ½ inch thick and as tender as the heart.

Any time you may have a failure in your garden, don't give up. Try to work out a method for your own type of soil and environment. Don't give up, be stubborn—oops!—I mean, persevere!

—Lucille Eisman

Second-Crop Celery

At the base of each individual stalk of celery, there is a tiny bud which ordinarily never sees the light of day while the plant is growing. But when the outer stalks are carefully cut off, these minute buds seize the opportunity to start growing.

When this happens the main celery should be used or stored as soon as large enough. Cut the "head" carefully, high enough to leave some sprawling outer leaves to support the roots and whatever "sidekicks" the plant may have. Select the best-looking shoot and cut out others at the heart so they can't grow. With the tremendous root system which is left in the ground, these shoots make very fast growth. It is like the rate at which a tree sapling grows compared to a seedling.

Where the season is long enough, the time of eating tender celery fresh from the garden is lengthened at both ends. Here in Florida it is done by "snitching" in the fall, and having a second crop in the spring. Further north it would be the other way round —taking early cuttings, as described, then encouraging the late-season growth by leaving a choice shoot to develop on the plant's strong root system.

—Donna Brunner

Home-Grown Celery All Winter Long

I had learned how to have carrots fresh from my garden all winter long by covering the patch with baled hay. It would be hard to find anything to equal the insulating power of packed hay or straw. If only I could use it to protect celery, too. Well, why not? Why couldn't I surround the celery with baled hay? My husband and I decided to try it. We placed hay bales along each side of the row with a bale to close off each end. More bales went across the top, shutting off access to frigid winter air.

We did the same with our late winter cabbage. Both kept fine. None of the vegetables wilted or shriveled, since they still had their roots in moist garden soil. All winter long all I had to do for a bunch of celery or a head of cabbage was lift off one of the top bales and cut what I wanted. No digging a pit, no dragging the vegetables into the basement to have them spoil before I could use them.

The hay bales served several purposes. After their job as winter protectors was over, I used them to mulch the garden for two or three summers. Finally they ended up worked into the soil as a source of humus.

Well-rotted manure can be applied liberally to the patch. Other excellent organic fertilizers include activated sludge, animal tankage, fish scrap, cottonseed meal, or dried blood, plus wood ashes, granite dust, or greensand for potassium.

Celery needs a lot of water because its roots do not go very deep into the soil. However, it is subject to fungus diseases, so if you must irrigate try to wet just the ground and not the stalks or leaves. Rotating your crops will also help prevent this problem. Try not to plant celery in the same spot for at least 3 years and give it as much sunshine as possible.

For indoor starting use rich garden soil and transplant at least once for stockier plants. Seedlings are rather slow and take about 10 to 12 weeks to grow to a size suitable for transplanting into the garden. They will withstand some frost and can be set out two or three weeks ahead of tomatoes.

—Lucille Shade

41

CHICORY

"Root" for Chicory

Unlike other leafy vegetables, witloof chicory is grown from the root of another vegetable. The seed is sown on well-prepared ground in May or early June in rows 18 inches apart. The leaves are somewhat like those of the dandelion and the roots like white carrots. In October, when the foliage begins to wilt, the roots are harvested. First, they are gently loosened, then pulled, put in heaps about 3 feet high, and well covered with a thick layer of soil. Do not remove the leaves when clamping as this can lead to premature growth. Where the weather is severe they should be stood close together in a trench covered with 5 or 6 inches of soil with a good layer of mulch on top to prevent freezing.

The best roots for forcing are those of medium size, about one to two inches in diameter; large roots tend to produce loose double heads. Thin side roots are cut off, the main root cut back to 6 inches and the leaves trimmed to within an inch of the crown.

Where only small quantities are needed they can be planted in deep boxes or large flower pots. Spread the soil to a depth of 3 or 4 inches in the bottom of the box and push in the roots in an upright position, allowing half an inch between the shoulders. Fill the spaces between with soil well pressed down with the fingers. Then give a good watering. No further watering will be necessary. If the soil settles, add more so that just the crowns are showing.

A further covering of 6 inches of fine, dry soil or peat should be placed over the crowns; keep this covering dry and free from any decaying matter or manure, as the chicons may be badly marked by rot. Do not use sand as this gets between the leaves.

Several roots can be forced in large flower pots with another pot of the same size for cover. The planting is the same as for boxes, but after watering no cover soil is necessary. However, the drainage hole must be covered as total darkness is essential. If they are even slightly green, the chicons will be bitter. It takes about 3 to 4 weeks for chicons to develop in a temperature of about 50 degrees. They can be forced at a lower temperature but it takes longer.

The chicons are ready when they are 5 to 6 inches high. They resemble miniature cos lettuce. They should be cut or twisted off the old roots and cleaned with a soft cloth, from the base to the tip. If the heads are not wanted by the time the tips push through the soil, another inch or so of covering should be placed over them. They should be used as soon as possible after cutting.

—Rosa James

Grow-It-Yourself Coffee

If you are growing chicory primarily as a beverage, order seeds of a "coffee" or "large-rooted" variety. We plant Giant Magdeburg. Salad varieties are usually intended for the production of blanched greens, widely used in Europe although low in vita-

I put the roasted chicory into a cloth bag such as a flour sack to keep the pieces from scattering, and pound it with a hammer to crush it. If the pieces are broken to coffee-bean size, they may be ground in a coffee mill, but I find the hammer method easier. Pound until most of the chicory resembles coarsely-ground coffee; a few larger pieces do not matter.

Chicory thus prepared can be made into a beverage in a percolator or by any other method used for coffee. Use about three-fourths as much chicory as you would coffee to produce a brew of comparable strength.

—Adele O. Laux

After washing thoroughly, cut roots into quarter-inch sections, slicing lengthwise.

mins. However, you can expect a bonus from you beverage chicory. The first young leaves, up to about 6 inches long, are good in salads without blanching and are often the earliest usable greens to appear in the garden.

Thin the plants 18 inches apart. We have obtained the equivalent of a pound of coffee from a single root when the plants were given ample room to reach their maximum size.

The roots are ready for roasting in the fall, beginning in late September. Wash them thoroughly and cut into sections not more than ¼ inch thick, it is "done" when brittle and dark brown all the way through.

Spread the slices on baking sheets and put in a 300-degree oven for about 4 hours. Chicory is cooked when it is dark and brittle.

COMFREY

Something for Everyone

Valuable in yield and versatile in use, comfrey may fairly be called "the plant with something for everyone." We serve it as a salad green and cook it as a vegetable. We use it as a mulch and for composting, and we esteem it highly for its beautiful foliage. Finally, it makes excellent fodder, and may be fed livestock green or dried, like hay, or made into silage.

Several years ago, at the end of October, we set out 24 crowns in a very small area which allowed each plant a space of only 12 inches. By the third week in November we had a hard freeze and the ground went solid until next spring.

Naturally, we thought we had lost all our plants, but 18 came through magnificently to compete with the first tulip tips. Growing with surprising speed and vigor, their vibrant green leaves shot up abundantly to start our education in the habits and uses of comfrey.

We soon learned that each plant needs 9 square feet of growing room in which to spread its leaves. Since our garden is small, we began to give away plants and, after disposing of 18 in 5 years, still have 17 producers. The taproot of a well-established plant may go down 10 or more feet in its search for food and water.

A perennial that can stand temperatures down to zero, comfrey does well in sweet soil with a hay or straw mulch, and requires plenty of organic nitrogen. In our experience, it has proved to be pest-resistant. Its many lance-shaped, slightly bristled leaves form a rosette, gracefully bending over concealed stems, and growing to a length of 20 inches or more, if not cut. When allowed to bloom, comfrey produces clusters of blue-lavender bells hanging on tall, hollow stems.

The yield of a mature comfrey plant after its third years sometimes appears fantastic. The plant should be cut several times a year, and by mid-May we prune all growth back down to two inches above the soil because the plant loses nutritive value if allowed to stand and bloom. By late fall, we may have cut the plant back 3 or even 4 times. If we neglect this pleasant and rewarding chore, the weight of the large, uncut leaves and stems can cause the massive growth to collapse.

Disposing of the harvest is no problem. It may be chopped up for a mulch, dug into the soil, or added to compost. Its high nitrogen content and potash make it valuable in the compost as an activator, and we find we can make a highly satisfactory finished compost by digging trenches 5 to 6 inches deep, and then covering the comfrey with soil.

We eat fresh comfrey from the time its leaves are two inches tall, and recommend it as a salad green whose flavor cannot be equaled, particularly when it is mixed with other greens in combination salads. One of our neighbors esteems it highly on bread, as a sandwich. Comfrey may be cooked like spinach, and we find its flavor much superior. However, cook it over a low flame—if heated over 125 degrees, comfrey loses much of its valuable vitamin content. It also makes delicious, healthful tea, brewed either from fresh or dried leaves.

—Hazel I. Diaz Pumara

Start Your Own Comfrey

Because of its deep root system, comfrey will improve land while giving bumper crops. Adelaide Peter of Marion County, Oregon, started with 25 plants in 1958 and now has more than 5,000 under cultivation.

In the spring, as soon as the season permits, she starts a new planting by running a power mower over the site. She then marks it into crisscrossing checkrows, each 3 feet apart, and, using a spade, works up large "pot holes" at the intersection of each row —about 4,500 plants to the acre.

Into each "pot hole" Miss Peter sets a crown cutting taken from a parent plant. The "start," which was taken from the two-inch level, is also planted at that depth. With frequent harvesting and age, comfrey plants develop many crowns. A crown-cutting is made by removing one of these with a sharp knife, cutting it off about two inches under ground. This "start" Miss Peter plants two inches deep in the readied "pot hole." The number of crown-cuttings removed from each plant depends on plant size.

Root-cuttings are taken in a different way from plants grown from leftover root and crown-cuttings. Miss Peter digs up the entire plant and removes most of the roots by breaking or cutting them int 2- to 6-inch pieces. These root-cuttings are also planted two inches deep, but laid flat in the planting hole.

It takes about a month for growth from such cuttings to show, and Miss Peter then removes the leaves from the plant tops on which she left some root, and replants them where they formerly grew, but about two inches deeper. Within a couple of years she again obtains root-cuttings from these plants.

After the plot is planted, she hauls fresh, very strawy manure from the dairy barn for mulch, applying it 3 or 4 inches deep over the entire area, and covering the newly-planted "pot holes," too. Later, as grass and weeds appear, more strawy manure is added. Otherwise, no more is done to the plot until the comfrey is ready to cut.

—Velma Davis

CORN

Snow on the Slope, Seed in the Pocket

Snow still clings to the nearby mountain ski slopes as Sam Ogden of Landgrove, Vermont, fills his empty right-hand pants pocket with corn seed. The time is May, and most danger of frost is past. Slowly, without bending over, he walks down the garden row, dropping 4 or 5 kernels in each hill, spaced 30 inches apart. And another corn season is begun.

"At each planting spot, I remove a shovelful of earth, fill the hole with compost or well-rotted manure, tamp it down, and with a broad-bladed hoe cover with half the amount of soil which was removed. Compost seems to produce better growth than does manure. Three stalks per hill are enough; excess plants can be pulled out."

After he drops the seeds, Ogden pulls the balance of the loose earth over them with the hoe and tamps it down so that the seed lies under half to one inch of firmed-down earth.

Ideal conditions for planting corn are in a warm, moist soil. In cold, wet ground the seed will rot, but with good seed and proper soil conditions, there should be nearly perfect germination and the shoots will break through in 5 days.

Warns Ogden: "Nothing will be gained by planting before the soil is just right; the seeds will take longer to germinate, and many of them will fail completely so that it may be necessary to replant. Cultivation should be done as soon as the shoots first mark the hills.

In Madison, Tennessee, spring is already well under way by the first of May. The clover cover crop is knee to waist high at H. L. Rushing's garden. He'll be out using a rotary plow with a mower attachment to shred the clover before plowing it into the soil about 6 inches deep. Next he'll use the plow to make furrows 4 feet apart; then he too will plant his corn in hills.

Rushing takes great care in building up his garden soil, especially before planting corn. Actually, he sows crimson clover over his entire one-acre Tennessee garden area in late summer. He rotary plows the soil in his corn patch at least 6 inches deep, laying off the corn rows 48 inches apart.

He applies a mixture of organic fertilizing materials in the row furrow in each hill spaced two feet apart; he then mixes the fertilizer with a half-inch of soil before planting seed. Now comes his favorite trick!

Drop 3 or 4 corn seeds and 4 pole bean seeds at each prepared hill site and cover one inch deep. When the seeds are up, the corn is thinned to two plants per hill. The hill distance spacing is used for Golden Bantam sweet corn and Hickory King white corn.

"If heavy rains occur prior to emergence of the corn and bean plants, we break the soil crust at each hill with a fork hoe or rake to assist sprouting

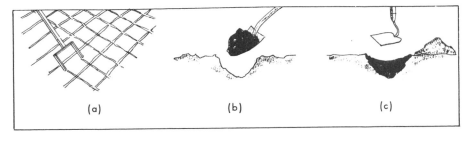

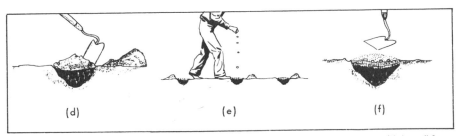

a)Mark plot off in 30-inch squares. b)Remove shovelful of earth. c)Fill with compost, tamp with hoe. d)Cover compost with half removed earth. e)Drop seeds. f)Hoe soil over seeds.

seed. If hills are spaced two feet apart in 4-foot rows, we've found pole beans do about as well planted in the corn hills as when planted adjacent to bean stakes. We just don't like the tedious task of staking pole beans. During hot, dry weather, the corn shades the tender bean plants."

When the corn is 6 inches high, Rushing hoes and thins the corn to two plants per hill. He tills with his rotary cultivator when the plants reach 12 inches, and at 24 inches he mulches with straw or wood chips. If a drought strikes about this time, he waters the corn and beans liberally about once a week.

Observes Rushing: "Needless to say, the profuse deep-green growth of corn and beans makes a wonderful sight. The corn worm is practically non-existent on the ears; the bean beetle may slightly attack bean plants when they are about a foot high, but then they disappear."

Rushing credits much of his corn-

growing success to the soil conditioning powers of his cover crop.

—Jerome Olds

Our Corn Ignores Drought

Although western Oregon sometimes seems overly blessed with rainy winters, the summers are usually droughty. And on our thin-soiled, rocky hillside that quickly dries out, a successful corn crop seemed impossible without irrigation.

The first spring after moving here to the foothills of the Cascades, I had hopefully dug out rocks and rotary-tilled a large plot, planting Reed's Yellow Dent field corn which we prefer to the sweeter sweet corn varieties. By the time the corn was shoulder high, the graceful green leaves began to curl and soon dried up on brown-parched stalks as the ground became "bone-dry." I harvested only fodder for the cattle. The second season was a dismal repeat of the first one.

The third year, however, I proudly watched the corn grow tall, sturdy and undaunted by the long dry period. "Why, this is corn like Grandfather used to raise!" exclaimed a relative in admiration. "How did you finally achieve this?"

"By following Grandfather's ideas," I confessed, knowing this would be the method I would follow hereafter.

Now, each fall, I spread fresh, very strawy cow manure—about 6 inches deep—on next year's corn plot. On top of this manure mulch, I pile a two or three-inch mulch of leaves.

In the spring "when the leaves on the oak trees are the size of squirrel's ears," I mark out my row and, using a triangular-headed weeding hoe, open a 5-inch-deep furrow. Then I drop the kernels of corn about 10 inches apart in the bottom of the furrow, covering the seed with an inch of soil. Lastly, I strew a light layer of compost in each of the furrows, spaced 3 feet apart.

When the corn is about a foot tall, I pull all the soil back into the furrow around the stalks. Then I mulch with old hay, straw, grass clippings, weeds, or whatever material is available. And I try to apply mulch weekly until the first roasting ears are ready.

Last year was the driest year of all, but our corn flourished.

—Velma Wilkinson

Six Inches for Corn

In rich, organic soil you can space corn as tightly as 6 inches both ways and get excellent results. When I ran a series of experiments to determine if a small kitchen garden could be made to pay for itself, I found that tremendous crops of sweet corn could be grown in very limited areas.

It's easy to plant seeds in small pockets or holes made in a permanent mulch of hay.

Look at it this way. Sweet corn is often grown in rows or "hills" in which 4 or 5 plants are crowded together within a few square inches, while several feet are left between them to allow room for tractors with their attachments. There is no good reason why a home gardener should follow such practices. Even if you just have a narrow area 25 feet long and about a foot wide you can plant sweet corn 6 inches apart both ways and get 3 rows of corn with 48 stalks to the row.

That is very close planting, and unless your soil is rich, you may have to add manure, bone meal or another supplement for best results. But my experience has been that you can count on at least 10 dozen ears of corn from that much space, although some of the ears will be a little on the small side and a few stalks may bear no edible corn at all. Over-all, however, you are getting remarkable returns.

Furthermore, you can use the site for other crops at the same time. Pumpkins and winter squashes are the best I have tried. Planted along the edges, they start well and, after you cut your cornstalks, spread over the whole area. So, there is really no such thing as having too little room for sweet corn. I recommend wider spacing if you can afford it, but excellent crops can be had in small spaces, even outside your regular garden, in "blocks" here and there around the yard. In my experiments with corn, I several times grew more than 6 dozen just by pushing seed through the mulch between my blueberry bushes along the front fence, and then waiting for results.

The best way I know to grow corn is in a permanent mulch. At planting time mark your rows with stakes and string, and push the seed through the mulch with your fingers. That may sound like hard work, but it is really very easy to do on a small scale. For best results, open a small "pocket" in your mulch for each seed, and press the seed into the soil at the bottom.

If you grow corn as I do in the same area each year, returning all husks and stalks to the soil, it will soon be hard to say just what your mulch consists of. Some organic gardeners also report that a second feeding steps up or prevents slowdown of corn growth.

In good organic soil, all corn plants should be dark green. And if they are, you have nothing to do to them but admire them while awaiting your harvest. A light-green or yellow color, however, indicates a lack of nitrogen, and you should take immediate steps to supply more. But do not apply any chemical fertilizer! A deficiency of nitrogen at this early stage is a sure sign of poor soil. And if you simply add quick nitrogen, you will be in for more trouble later on. What your soil needs is an organic fertilizer like animal manure. You can hardly over-do such application on your corn. Manure will not "burn" it, and whatever is not absorbed by the current crop will enrich the soil for the following year. Once more, as you can see, I am assuming that you will go along with me on "crop rotation" at least to the extent of growing your corn in the same place for a few years. Try a few locations, choose the best, and then *stay* with it while you improve it.

—Richard V. Clemence

Corn All Season Long

Out here in northwest Wisconsin we pick our corn every day for the table because we plant Sugar and Gold, a 57-day hybrid; Extra-Early Golden Bantam, a 69-day corn; Golden Bouquet, a 77-day hybrid; and Cheddar Cross, an 88-day hybrid. We plant all these varieties at once, letting them get off to a good start together.

Planting a packet each of these varieties helps spread the canning season, although I personally favor Sugar and Gold and Extra-Early Golden Bantam because I have found that corn is sweetest when the weather is hottest. Notice that they mature about 10 days apart, which is just right for our region. When we tried working with 3 varieties, we missed out on having a continuous supply.

—C. F. Enden

Start from the Ground Up

The soil we grow our corn on is a heavy, fine-clay loam which had just been cleared when we started. It needed some building up because it was on the acid side, so we added lime and bone meal, with a good sprinkling of ashes, to bring the pH reading between 6.0 and 7.0. We worked these soil conditioners in with that handy garden power tool, the rotary tiller—and then spread heaps of cow manure that had a lot of hay bedding mixed in. Everything was then plowed under, and the ground was smoothed and marked for planting.

We plant as soon as the soil is warm in the spring about 1½ inches deep, in rows 18 inches apart, with 30 inches between the rows. The corn is grouped in several rows of one variety to encourage good pollination when the pollen falls from the tassels onto the silks. This good start is passed to the cob when the kernels begin to form.

It usually takes about a week for the corn to mature after the silks get brown. When the silks are dry and brown, and the ears are round and full, the corn is usually ready to use. Golden Bantam can fool you because the ears stay so slender, even when ready to eat, that you can hardly believe they are ripe. We have found that we have to watch it more closely than others.

Corn needs plenty of moisture, plus good drainage. It especially needs a good supply of moisture when the cobs and kernels are forming, and we sometimes have to irrigate in a dry season to save our crop. A layer of straw put on when the weather gets hot helps hold moisture for the developing corn.

The organic treatment we give our corn shows in its rich green foliage. We encourage the feeder roots to grow out into space between the rows where we turn under plenty of nutrients—cottonseed meal, ground phosphate rock and greensand. Once when a punishing rainstorm knocked the stalks down, they soon straightened up because their sturdy root systems held firm. The late varieties stretch up to 7 feet, but the early-bearers are on short, small stalks. We think so much of our corn feeder roots that we don't hoe too deeply—we're afraid of damaging them.

Before I pick corn for canning, I first get all my canning materials ready so I can work quickly and easily while the corn is at its peak. I find that *the fresher it is picked and used —the sweeter and better it tastes.* After picking and shucking it, I wash and rub the ears in a large container of cold, fresh water which gives them "tone," and removes the dust and excess silk. I then cut off the extreme tips of the kernels, scrape the pulp out, and add enough water to obtain just the right thickness or consistency, heat carefully and cook.

The amount of water to be added varies with the maturity of the corn, and it takes a little practice to get it just right. When I freeze corn, I first cook it, cool it, and then cut off the kernels. When the corn doesn't cut and scrape easily, it is too hard to use.

—C. F. Enden

Grow the Earliest Corn

Sweet corn is such a delight to our family, it always seems a shame we have to wait until July to enjoy our own. So we decided to do something about changing that in our area near Philadelphia, Pa.

First of all, we needed to start with a really early-maturing variety. There were quite a few, we found, that were ready to eat in about 60 days after planting—Seneca 60 and Seneca Explorer, Early Sunglow, Spancross, Early Giant Golden Bantam, to name a few. We settled on Early Sunglow: advertised as the earliest of all, and in our opinion, as good as any sweet corn we'd ever tried.

Corrugated plastic bends readily into a "U" and can be tied securely with pliant wire.

Early Sunglow is a fooler. When we first grew it, we thought the plants were suffering from some kind of soil deficiency because the leaves had a decidedly purplish hue. Later we found out this was its natural color. We were also fooled by the size of the ears. They seemed so small on the stalk that we figured the kernels would have to be extremely shallow. But to our surprise, the ears had such small cobs and so little outer husk, that they actually had about as much corn on them as some later-maturing varieties. And the kernels were deep enough for a good, juicy biteful.

We began to push planting dates up earlier. That's another advantage of the short-season corns. They will take cold, wet spring weather better than longer-maturing varieties. I finally tried planting in March, like the big Midwest commercial corn growers are doing. You can get away with that on light, sandy soils, but with an especially cold spring, I

didn't have much luck on my heavier land. Unusually cold and wet weather delayed germination, and in fact, about half the corn didn't come up at all. What did push through the surface of the soil just sort of stood there and turned blue in the cold weather. Some died, while some fell easy prey to birds, slugs and other pests. But what did survive really shot ahead when warm weather finally came, and we had a few ears before July 4th—unusual in my garden anyway.

Next I tried starting corn in a cold frame, then transplanting it to the garden when the soil was fit to work. But corn doesn't transplant very well, I found. So I grew some in peat pots in the house and set the pots out when the soil warmed up. This worked all right, but it certainly wasn't practical. It takes a lot of time and peat pots to set out a decent-sized patch of corn. And even then, I still had the birds—who seem to like that earliest corn best—to worry about.

Place shaped and tied plastic panels over rows, overlapping them for full coverage.

I was ready to give up on growing corn any earlier. Then one day when I was trying to figure out what I should do with a stack of corrugated clear-plastic panels left by the former owner of our home, I got an idea. Why not bend the panels to make covers over the rows of my early corn? The plastic would make a sort of miniature greenhouse over the plants, keeping out cold air but letting sunlight flood in to warm the soil. Furthermore, the panels would keep out pesky crows, blackbirds and rabbits.

The idea worked like a charm. I merely bent the panels to the desired form, tied them in that shape with two loops of string to keep them from springing back flat, then set them over the row with stakes on both sides so the wind couldn't blow them away. No more bird worries—I could look at a crow flying over and actually smile!

Once the corn was up under the plastic covers, it grew reasonably well

Boards at the ends of tunnels keep rabbits out and the heat within during cold nights.

for cold weather. But I found that by mistake I had planted a middle-season variety instead of Early Sun-glow in the excitement of experimenting with the row covers! So we didn't set any records for earliness that year.

Incidentally, plastic row covers aren't my original idea by any means. Southern California has whole acres planted under them, and both Rutgers and Cornell have experimented with the idea, especially for tomatoes and other tender vegetables. But in these cases, growers use plastic *film* rather than semi-rigid panels, stretching it over various kinds of wire frames. Panels are considered too expensive for large-scale commercial projects. But for the home gardener —who needs only a few panels—it's worth trying if you like corn as much as we do, or if you are having a particularly difficult problem with birds or rabbits.

But even under the row covers, I noticed that the corn had taken a

To prevent tunnels from blowing away, bend steel rods before thrusting them into soil.

long time to germinate because of cold soil. So this past year, I tried still another trick to speed up growing time.

I soaked the corn seed in warm water for about 24 hours before planting. The seed swelled in the water and seemed just about ready to sprout when I put it in the ground April 13, planting shallow to take advantage of whatever warmth had penetrated into the soil that early. Most of my crop was up and growing under the row covers by April 22, ahead of all my gardening rivals. By May 4, some plants had reached five inches high, and we got ready to enjoy corn on the first day of summer.

Now I am already looking to the new season—and the possibility of shaving a week to 10 days more off my schedule for earliest corn. First of all, I'll plant in March, in rows that I will have previously covered with my plastic panels to warm the soil a little more than normal at least. We usually get a warm spell toward the latter part of March, and if we catch it right, there's usually a day when the ground is dry enough to work up. (Anyone with sandy soil in this area ought to be able to stay a week ahead of me.)

Then I will pre-soak the seed until it actually sprouts before I plant it. I'll remove the row covers, plant the seed, cover it gently so as not to break off the tender sprouts, and replace the covers. With a little luck, we may be eating fresh corn this year around June 10—nearly a month earlier than normal!

—Gene Logsdon

Three-Course Feast

In a radio broadcast one morning we heard the expert say that sweet corn shouldn't be mulched, since it might delay and even prevent its maturing, especially if the summer happened to be cooler than usual. Joe and I looked at each other with the same disconcerting thought uppermost in our minds. We had mulched our biggest plot of sweet corn *because* we were having a cool, wet season, with everything getting off to a bad start. The corn, we figured, might get an extra push from some added soil food and a covering of straw. And it needed that push because it was only three inches tall by the Fourth of July.

The corn had been planted on ground where tall green sunflowers had been chopped into the soil a year earlier, and the earth was good. But as the first booster we applied a layer of compost when the stalks were about 18 inches high. When the corn had grown another foot, we added a layer of duck manure in a wood chip litter; and a foot later we covered everything with about three inches of straw.

To make a real comparative experiment of this, we left two other sweet corn plots without the booster feedings; and it wasn't long until the differences showed. The mulched plot outgrew the others, both up and across; and its color was several shades darker, deeper green, with broader leaves and stalks at least one-third thicker. The corn stopped growing up at a height of 8 feet, compared to 5- to 6-feet heights on the comparative plots. And there wasn't a weed to be seen in the test plot, while we fought the pesky things all summer on the other patches, with the cool, wet season allied against us.

But the high spot of the whole experiment was the food quality. Our "boosted" corn had huge ears filled to the very ends and the deepest golden-yellow kernels we ever had seen. And the flavor was unlike any sweet corn we ever had raised before.

—Ellen Perry

Savor the Flavor—of *Roasted* Corn

The finest of all corn flavor is obtained from *roasted* corn, not steamed or boiled. Making a corn roast is a bit of a production, but it is well worth the effort once or twice a year to produce a taste thrill that is so glorious mere words can not describe it. Some people roast corn on a charcoal grill by wrapping the ears in aluminum foil, but it is difficult to produce enough quantity by that method and often you can't get enough heat to do the job properly. I admit that my method will burn a hole in your lawn, so it is recommended that you set up your fire on a vacant lot or gravel surface, but after you taste the end product you won't mind that extra inconvenience.

Here is what you need:

1. A sheet of steel ⅛ to ¼ inch thick and at least 3 by 3 or 3 by 4 feet in size.

2. 6 or 8 concrete blocks.

3. 6 to 12 burlap bags, depending on how much corn you want to roast.

4. A sprinkling can and some water.

5. Plenty of good, dry firewood, preferably the kind that will produce plenty of hot coals.

6. Finally, a bushel or more of fresh-picked corn in the husks.

Step one is to set up your concrete blocks so that they will support the steel plate about 18 inches off the ground. I usually set the blocks in two rows, with each end open and perhaps a small gap in the middle of each side. Please note that this whole method is very informal and there is nothing tricky to worry about. The only factor that is absolutely necessary is plenty of heat and steam.

Set the sheet on the 18-inch-high blocks.

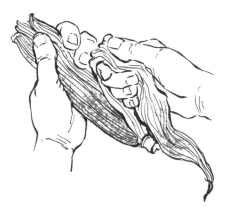
Leave 3 to 4 leaves when stripping corn.

Set corn on the sheet when the fire is hot.

Wet burlap and sprinkling prevent charring.

Don't put the steel plate in place on top of the blocks until after you have built a good fire. In fact, keep the fire going for perhaps one-half to three-quarters of an hour in order to build up a good bed of very hot coals. While you are getting the fire hot you can also be wetting down all of the burlap bags, and stripping the outer husks from the corn. But be sure to leave in place about half of the husks, the inner 3 or 4 "leaves."

Then put the steel plate in place and lay on top of it two wet bags. Next, cover this cooking surface with corn ears laid side by side, right up against one another. Then place two more wet bags on top of the corn, and if you have more corn to cook make another layer of ears. Finally, cover the top layer of corn with burlap again (3 or 4 bags this time). If your fire is hot enough the whole corn roast should start sizzling and steaming almost at once. Whenever you sense that the whole mass is going to burst into flame or start charring too much on the bottom, sprinkle it with the watering can. Keep stoking the fire as needed.

Just how soon the corn can be taken out and eaten depends on the heat of your fire and the construction of your fireplace. Usually you can figure on one-half to three-quarters of an hour. The most common mistake is to take the corn out too soon. When finished, the bottom row should be charred on the bottom and the husks of all the ears should look to be well cooked. What gives roasted corn its special flavor is the slight taste of the husk and of charring that permeates it. If you're a devotee of steamed corn you may wonder how any improvement could be made on the taste, but one mouthful of roasted corn will convince you that it is better.

—Robert Rodale

Put Corn in the Pot

For plenty of sweet corn in the pot until time of heavy frost, new plantings of my favorite late variety, Iochief, are made every two weeks or so until the latter part of June. The first and last plantings are always a gamble as to whether the seed will germinate or the kernels fill out in time. That is how much and how long my family enjoys eating sweet corn.

As each planting of corn reaches 10 or 12 inches, it is thinned out to 10 or 12 inches apart. Bone meal or compost is then sprinkled along each side of the row to a width of—again—10 to 12 inches, and the planting area cultivated, working the rotary tiller up close to each side of the cornstalks. Using a rake, the loose topsoil is pulled over and up against each corn row from each side, "hilling" it up 6 or 8 inches high.

Next, bales of old or spoiled hay are shaken out in the area to make a continuous mulch 5 to 6 inches deep. It is always wise to work with the wind at

Thick bats of mulch are easily peeled off hay bales and laid on soil between planting rows. Hay not only conserves valuable moisture, it also builds soil fertility.

your back to keep the dust and seeds away. Sometimes the hay separates quickly and easily into one-inch pads or flakes which are equal to 5 or 6 inches of shaken hay.

The continuous, between-row mulch of loosened hay is spread right up to the hilled-up corn to make walking through the patch a pleasure, while it lessens moisture evaporation and keeps the soil underneath cooler during the summer heat. It also adds nutrients to the topsoil while decomposing, and gives more tilth and sponginess to the entire garden at rotary-tilling time.

At this time the corn is "laid by" as the old-timers say, with no more hoeing or cultivating except the pulling of an occasional weed or two between the cornstalks. Constant watch is kept for smut on the ears or stalks; when found, it is quickly picked and gotten rid of by burying, composting or garbage disposal. This fungus growth can spread to other stalks, ears or carry over in the ground, so I battle it right down to the last ear, pulling each cornstalk that has finished bearing and remove them from the garden. I believe improper air circulation has some bearing on the smut problem.

—Edward P. Morris

Mineral Oil for Beetle Foil

When the silks appear on the ears and the inner ear is about half formed, a squirt of pure mineral oil into the silks will foil black beetles or corn earworm moths. I use a pump-type oil can with a long spout, having found that mineral oil seals the silks together and protects the end of the ear against the penetration while in no manner impairing the flavor. Later there is a reaction of the mineral oil on the silks, causing them to come off easily with the husks, leaving the corn practically free of the silks usually left between the rows of kernels. The few remaining are quickly removed by the light use of a vegetable brush.

—Edward P. Morris

Radio Keeps Out the Raccoons

I grew sweet corn for sale on the edge of our small town. As soon as the corn was ready to eat, the raccoons moved in. A friend told me to put a radio in the patch. It worked like a charm. As long as the radio was on all night, the raccoons never bothered the plants. But leave it off a night or two—and they were right back. Be sure it's a station that's on all night. If the garden is too far from a building to use an extension cord, use a battery-operated radio. I turn mine on at dusk, shut it off in the morning. I am sure this advice will benefit many readers.

—Sam Hughes

Aluminum Foil for Better Corn Production

Aluminum foil can increase corn production 15 percent or more by reflecting light on the underside of stalks. A series of tests conducted by the College Experiment Station, Athens, Georgia, and reported by Associate Agronomist H. F. Perkins, showed an "increased corn yield of 22 bushels to the acre," while lodging fell from a high of 36 percent to 7, and barren stalks fell from 4 to less than one percent!

Sharply increased photosynthesis—better than two to one—is held responsible for the higher yield and better plant health. Strips of ordinary crinkled aluminum foil—again like those you buy in any market—were strung between rows 3½ feet above the ground. A total of 37½ square feet of foil was exposed on each side of a 20-foot row. A "control" row right next to it was treated in exactly the same way, but without the aluminum foil reflectors.

—Maurice Franz

"Nightshirts" for Sweet Corn

Bags of any kind—paper, cloth or plastic—make effective "nightshirt" protection for late-ripening sweet corn when overnight thermometer dips are expected. So do plastic or ordinary bed sheets, which Colorado gardener Dorothy Schroeder says she quickly fastens down with clip clothes pins, leaving them in place when there's a succession of chilly nights. Plastic bags or sheets also work well over peppers and eggplants, she adds, and "you can even cover a grapevine with a couple of the sheets, using a dozen or so pins to hold them in place."

—M. C. Goldman

CUCUMBER

On A Trellis or Fence

If you've kept yourself from growing cucumbers because you lack space in your garden, cheer up. By growing cucumbers on a fence or tellis, any gardener can grow and reap a bountiful harvest in less than one-fifth the room ordinarily required. If you have space enough for another row of carrots or beets, you have room enough for a row of cucumbers!

One row of about 20 plants produces more than enough for an average family. When grown on a fence, cucumber vines produce like mad because they have abundant sunlight and a good circulation of air that prevents disease. And, too, bees have better access to the blooms, thus pollinating far more than under ordinary curcumstances.

There is no specific climbing cucumber. All cucumbers have a tendency to climb. Just choose your favorite all-around cucumber, and go ahead. We have found it wise to plant two varieties—one just for slicing, and one for pickling. If you wish, one variety will take care of both needs.

Manure has been found to be the best food for cucumbers. If it is scarce, mix a small amount into each hill, cover with soil, then plant the seed. But if enough is available, turn it into the entire area at least three weeks before planting time.

If no manure is available, an all-organic fertilizer (such as a 5-10-5) equivalent in NPK prepared from cottonseed meal, rock phosphate and granite dust may be broadcast prior to planting. Or, add a small amount to each hill when planting.

It is best to work and fertilize the row before putting up the fence, since the soil should be loosened thoroughly down to 12 inches.

The material that cucumbers like best to climb upon is 20-guage poultry wire. The thin wire provides good anchorage for the springy tentacles. The fence should be at least six feet tall. If you're unable to get wire that wide, overlap two narrower strips. Since the vines produce mostly on their tips as they grow taller, it is important to give them plenty of height in order to prolong the producing season.

Before planting, place seven or eight-foot poles every four feet along the row. Hammer them securely into the ground, then nail the fence to them, stretching it tightly between the poles. Do not space the poles any further than four feet apart, for the vines become quite heavy with fruit by midsummer, and will cause the fence to sag if sufficient support is not present. Be sure to place a guide wire on either end of the fence to prevent sagging in the middle.

If possible, run the fence from east to west. This will keep all the vines growing on the south side. If the fence must be run the other way, try to keep the vines growing on one side for easier picking by removing the vine tips, now and then from the holes, should they grow through.

If your soil has poor drainage, plant the seed on a long, raised hill at least ten inches high. If drainage is good, just plant in a row as you would carrots. Plant six inches away from the wire, and place six to eight seeds every two feet in the row. Cover with a half-inch of soil, and firm gently with your foot.

When sprouts appear, thin out to stand four per hill. When six inches high, thin out to stand only two plants per hill. One word of caution here—the roots of cucumbers are usually tangled so that when a plant is pulled up, it disturbs the others. Either cut off the plants at ground level when thinning, or pinch them off with your nails.

Cucumbers may be given a side dressing of manure when the vines begin to climb. Or, the 5-10-5 all-organic fertilizer may be used. One side dressing is enough in early summer.

As for cultivation, cucumber roots are many and shallow. Cultivate only to keep down weeds, then hill high with soil taken from between the rows. Be careful not to hoe too near the roots. Hilling should be done a week after the side dressing has been applied.

After hilling, mulch the entire area well with grass clippings, old hay, or straw to preserve moisture and to keep down weeds. Train vines on the wires as they grow taller. If weather is dry, water down well once a week with an over-head sprinkler or ground soaker. Allow ground to dry out between waterings.

Cucumbers may be picked and eaten at any stage, from the tiny one-inch size to the huge, ripe cucumber. For pickling and relishes, use them up to five inches long. Above that, the cucumbers may be dilled, used in recipes calling for ripe cucumbers, or simply sliced for the table. Watch your cucumbers carefully from day to day as they grow very rapidly. Should you miss a few that turn yellow, keep them picked to stimulate further production.

To really enjoy cucumbers at their best, do not pick them until the moment you are ready to use them. If picked too far in advance they shrivel and lose flavor. And to grow them trouble-free and in far less space, get 'em up on a trellis!

—Betty Brinhart

The First Cucumber Tree in Texas

Maybe Texas is famous for "wide-open" spaces, but my Houston garden is quite small—and crowded. It was obvious that the only way cucumbers would fit in was to go—up.

But where could I put them—and how? There was no room on the fences and the tomatoes took the only trellis available. Then I spied a 6-by-8-foot piece of reinforcing wire and found a place down at the end of the garden path where I could set a cylinder and grow cucumbers around it.

So I made a perfect cylinder, 6 feet tall and 30 inches wide, rolling the wire and bending its ends to hook them firmly together—strong enough for anything. I lugged it down the garden path, set it in place and drove stakes on either side to secure it to the ground. Seeds were sown outside the circle, an inch or two from the wire, and about 3 inches apart. Almost before I knew it, little twin leaves were unfolding through the soft earth and my spicy chuck pickles were headed toward reality. When the plants were about 6 inches tall I pinched out every other one, leaving the strongest to climb the first cucumber tree in Texas.

CUCUMBER

16 cucumber plants flourish on a trellis covering an area less than 3 feet square.

Although our garden has grown, I still use the cylinder method, planning in advance where it is to go next, and preparing the soil. Cucumbers like a light loam, neither too much sand nor too much clay, with plenty of humus to keep the moisture around their roots. They like their soil stirred deeply so their roots can stretch like sleepy giants; and though cucumbers are a thirsty breed and demand lots of water, they will not flourish if their feet are wet. So I am careful to place the cylinder in a spot that is properly drained.

To be sure that my cucumbers have plenty of water, I dig a hole in the center of the circle before placing the cylinder, and sink a gallon metal can up to the rim, filling it with dried manure from the nursery—none other is available. At least every other day the can is filled with water which seeps slowly through the manure and out a small hole in the bottom of the can, thus feeding the plants and quenching their thirst at the same time.

There are other advantages to using the circular trellis method. If drainage is a problem, the cylinder can be placed on a raised bed. When soil needs to be amended for just the right amount of clay and sand, the small area requires less work than a long, wide row.

In rows, cucumbers should be thinned to about 12 inches, whereas with the circular trellis cucumbers can grow as close as 6 inches. In just a short time, with a little guidance, the lush plants will cover the reinforcement wire, and the tower of green will be a thing of beauty in your garden.

The fruit will be clean, there'll be no sunscald or white streaks as is sometimes found on cucumbers that grow upon the ground. For some reason—maybe to hide from our Texas sun—the cucumbers usually grow *inside* the cylinder, but since the reinforcement wire is made in 6-inch squares the fruit is easily picked.

—Marian L. Coonse

Fence—Planted Cucumbers Thrive on Electroculture

I've been practicing electroculture for the past several years with my cucumbers without actually knowing it. I just knew that I wanted homegrown cucumbers. I also knew I couldn't afford the space for them.

Well—our "postage-stamp" organic vegetable garden (12 by 18 feet) was fenced in, and even more fortunately the fence was a metal one. As supports, my husband had used some old metal pipe.

Last year, with only 6 plants growing along and on the fence we harvested about 200 cucumbers. What I did notice was that our greatest yields were realized after each rain storm. About 20 cucumbers would seem to suddenly appear, ready for picking. It was as though a magic wand had been waved—and presto —cucumbers galore!

After raising cucumbers in this manner I wouldn't switch, even if I had more space. There are ever so many advantages to fence planting. The beautiful blossoms are readily exposed and they attract many more bees. The fruit hangs from the plants and is a rich deep-green color all around.

Pea plants also occupy a portion of the metal fence and do very well. Lack of space needn't prevent a gardener from considering cucumbers. First—"plant" a fence (make it a metal one), then plant the rambling vegetables.

—Violet E. Simon

Staking My Reputation

My garden is frequently referred to as "Mrs. Taylor's stake-and-tin-can garden" and I might as well add a little to its reputation.

I use stakes because I like to know where things are planted and often have a dozen or more lines of string —binding twine—strung across so I, or anyone else, may run the little rotary cultivator and be guided the entire length of the row, even when the vegetables have not shown. There

may be a million tiny weeds just showing and that is the easiest time to eradicate them.

Tin cans, as I have used them for many years, consisted of rusty #10 cans partially set in the soil, one end open and the bottom out, or with many holes, then filled with fresh horse manure—when I can get it. I plant my vine crops: cucumbers, melons and squash. Later the vines will be mulched and all watering done through cans. I certainly want the fun not only of a new experiment for me, but also the fun of hearing the comments of others.

The tin-can use seems perfectly clear and I plan to plant some with more than one vegetable.

Our seasons here in South Dakota are always later than further east and south, but this year later than normal and since the 30-day forecast predicts still below-normal temperatures, we will need to use every opportunity to assist nature.

We know gardens seem to grow noticeably after an electric storm which releases certain elements in the air—one can often actually smell it— so why not attract the beneficial components all during the growing periods.

—Mrs. Irma W. Taylor

GHERKINS

Cucumbers That'll "Like You"

I became tired of explaining that "I like cucumbers, but they don't like me."

If you've got the same problem plus the desire to experiment I think you'll enjoy growing — and eating — gherkins and garden lemon or yellow cucumbers. Both varieties are delicious served fresh and crisp from the garden, and both make perfect pickles — the dill, the sweet and the sour — that have a truly tangy quality.

For a number of years now I have been planting a few hills to gherkins and the lemon cucumber. These somewhat neglected members of the Cucumis family require the same culture and soil conditions as their more conventional cousins. Give them a loamy, well-drained garden site with plenty of humus, compost and, if possible, well-rotted manure.

I plant in hills allowing them plenty of room to grow and placing a layer of compost or manure on the bottom of each hole before setting the seed. When the seedlings are about 6 inches tall, choose the 3 strongest vines and pinch out the others. Gherkins are very prolific and I find that 3 vines provide enough fruit for the average family. If picked regularly, they will continue to produce throughout the season—which begins in my area about 60 days after planting time in early June.

West Indian Gherkins have long been my favorite. As the sketches show, they bear a burr-like fruit which has given rise to still another name—Bur Cucumber. They develop a fat, oblong shape and although they can attain a two-inch length, don't let them get too big because they tend to go seedy and the skin is too tough for good eating.

If you're looking for a rare treat, here's my advice. Pick your gherkins when they're one-half to three-quarters of an inch long and serve them at the table 30 minutes later. You'll also find that, picked young and tender, they make the best pickles for after-season and over-the-winter snacks and meals.

Almost as versatile and just as good to eat, the garden lemon or yellow cucumber is another fat, short member of the family. When young—two inches long with a plump 3-inch girth—they taste like young, very tender, cucumbers. The fruit, when matured, is deep yellow with a pulpy inside that tastes like a cucumber crossed with a cantaloupe. At any stage they make fine pickles and are excellent preserves when picked dead-ripe.

If you're conservative in the menu department, just follow your own pickling recipes but use gherkins instead of cucumbers. Or perhaps you're really from Missouri and have to be shown even when your family is clamoring for new taste treats. If so, try this: Make a table relish of onions, green sweet peppers and green tomatoes, all chopped up with gherkins.

When you set this dish out on the dinner table you and your family will all be happy about gherkins, these "different" cucumbers from the tropics which can be successfully grown up North in our more rigorous climate.

Cucumbers in Compost— and Burpless to Boot!

There's a practical way to make your compost set-up do double duty. And for gardeners limited in space, it's an attractive and doubly valuable way to include some rampant-growing vegetables such as squash and cucumbers in a place other than the middle of a tiny plot.

Even if your compost operation—like mine—is a small one, it probably has two pits, one for the material ready to use and the other one for "curing." I utilize the second type of compost heap to grow my cucumbers to the very best advantage. If I try to grow cukes in my regular beds, they do rather poorly, as my garden is almost pure sand and only 25 feet away from ocean tide water. There's no depth to the soil and it's soon dried.

So, for the past few years, I have planted my cucumbers on top of my newest compost heap, where the roots go down into the heat and moisture of the developing compost and where, as shown in the photo, I can construct a trellis for the luxuriant growth to climb on.

If the pit is not available, I then plant the vines in an unused cold frame, where they can be treated to extra water and extra food, but where, quite frankly, they do not do as well as in the pit.

For this kind of growth, I plant an old cucumber variety called China, easily found in most seed catalogs, which when grown on a trellis produces 15-inch-long, tender-skinned, burpless cukes.

On the subject of being burpless, I have known for some 50 years that the secret seems to be mostly a matter of fast growth. If one compares the southern-grown outdoor cucumbers

Growing in a compost pit, facing the sun and protected from the wind, these trellised cucumbers rate high in food value.

sold in stores with the hot-house cucumbers, you can easily see how much more digestible the quickly-grown types are. In other words, if you want good cucumbers that never come burping back, remember that they need rich deep soil, lots of water, and heat enough to make them grow fast.

And one way to make a start on this is to plant cucumbers in your compost heap. Staking and tying them takes only a few minutes twice a week and ensures the fruit staying nice and straight. But if you have plenty of room on the horizontal, the plants will grow and bear just as well on the ground—though the cucumbers are more apt to be curved and irregular.

—Nelson Coon

We Had Cucumbers All Summer Long

Before sowing cucumbers, we loosened the hill with a spading fork to aerate the soil, then spread a bushel of old compost over the area. Next we scattered a cup of balanced mineral soil-conditioner on top and raked it in, leveling the hill. We sowed 6 or 7 seeds about 8 inches apart, pressing them less than an inch into the ground. A few large flat flagstones were placed on the hill to hold moisture and to keep the cats from digging there. (The 3-foot-high wire fence enclosing our garden keeps dogs and rabbits out, but cats scramble up and over to make merry in newly-sowed beds.) Boards or heavy cardboard will serve the same purpose, as will flat grape boxes. We remove the flagstones within two days, but grape boxes can be turned bottoms up and left two more days unless it rains.

About the same time we sowed another hill in a low spot at the west

side of the garden, which is shaded by pines after two in the afternoon.

When each seedling had two big leaves we thinned out all but 3 plants in each hill. After hand pulling a few weeds we mulched around the cucumbers with grass clippings. Then, except for the addition of more mulch a few weeks later—especially in the area where the vines were extending—the hills needed no more attention until picking time. We didn't even water during the long drought.

That south hill bore earlier and produced twice as many long, crisp cucumbers as the one on the west side, which we could have done without, for the south hill gave us more than enough cucumbers, day after day, for lunch, dinner and horse-radish pickles. It bore till the first severe frost. After that only the vines that had climbed the fence continued to bear.

One Rhode Island organic gardener near us mulches with thick layers of newspaper, putting hay, grass clippings or pine needles on top for a better appearance. And a southeastern Massachusetts gardener who believes in allowing each plant ample space for air circulation says, "Keep cucumbers picked and they'll bear heavy and keep bearing." However, it's easy to miss those big cukes hidden under dense, deep-green leaves; you have to feel around every day, and even then a few are likely to elude you.

—Devon Reay

As Large as a Loaf of Bread

Up at Schenectady, New York, cukes as large as a loaf of bread rewarded Hans Lueck's gardening skill. Many of them weighed 5 pounds and over, he relates and he had harvested 200 cucumbers from his 5 plants by the first of October. During the growing period, Lueck fertilized with cottonseed meal, dried blood and phosphate rock, worked into the soil and watered down with one tablespoon of fish emulsion in two gallons of water. His vines shot up 6 to 8 inches a day, he notes, and reached more than 15 feet high, making it necessary for him to use a stepladder to harvest and care for them.

And at Louisburg, Missouri—a state where folks have to be shown—first-year gardener Paul H. Robinson tried a test with his cucumbers to prove the Ruth Stout mulch method he had been reading about. Using an 8-inch blanket of spoiled hay between his vegetable rows, plus a plastic-hose system of delivering manure-tea irrigation to his plants, Robinson kept a careful record of the yield from his cucumber vines. The seed packet stated that the average yield should be 180 per 100 feet of row. After planting on July 10, Robinson started gathering ripe, full-sized cucumbers in the first week of September. By the end of the month he had harvested 2,166 of them—344 pounds in all—and was still reaping cukes. His crop return bonanza was about 12 times the suggested yield.

—Nancy Bubel

DANDELION

A Lover of the Greens

Being a dandelion greens' lover and regretting that they last only a couple of weeks in early spring in the wild state, I bought seeds from a nursery some 3 or 4 years ago and planted them in the garden. They are still there and in the same rows where they were planted, as I don't plow or till up the garden any more.

Each spring I loosen the soil and plant around them, then harvest the greens all spring, summer and fall. Besides, I always freeze the winter's supply out of probably 25 plants in all. Last summer they self-seeded a few plants, which I'll take up and put in the row this spring.

All summer, I dump the dry garbage and wastes right on top of the leaf and straw mulches, whereas the smelly type of garbage is hidden underneath. My earthworm population is sufficient to take care of such practices—and a good thing too, as I don't have the physical strength to do much composting.

The garden has had generous applications of rock phosphate, seaweed, wood ashes. When the soil became covered with moss from two many oak leaves, we put fresh marl over the entire surface, which corrected this sour condition.

The dandelions, being untemperamental and undemanding, get no ex-

65

tra care whatever and last summer they grew to a terrific size. I don't know whether the seaweed, the marl, the leaf mulches, the rock dusts or what accounted for this outlandish growth. They were tender and good-flavored all summer, whereas before, during the hot, dry weather, they were tough and I left them alone until the fall rains caused new, tender growth, after which I cooked them all autumn.

I dig the root vegetables from under the mulch all winter. The dandelions and cress will be much earlier this spring for having had this coverlet of leaves and straw through the zero weather. I uncover them by degrees and will probably find a mess of dandelions ready under the mulch in early April, possibly in March.

—Gertrude Springer

DASHEEN

Staple of the Tropics

When we started to garden here in Florida 3 years ago, we made a thorough "question campaign" among old-timers to locate seed. The answer we usually got was "Oh yes, we used to raise dasheen when I was a kid. But I haven't seen any for years!" Often we detected a twinge of nostalgia. Although commonly grown a generation ago, and especially liked by this area's Polish settlers, the earthy and humble dasheen has given way to supermarket potatoes and packaged foods. Finally we located a grower and bought a bushel of the small, brown, papery-covered bulbs—our start of dasheen seed.

One of the main reasons for the decline of this very adaptable and valuable food is a lack of soil to which it is suited. It loves deep, moist humus, with good drainage. Plenty of muck and mulch, which are scarce in much

of Florida and the Southeast, not chemicals alone, will give good results with this crop. Highly soluble plant food is not what it craves, but rather humus, moisture, sun and lots of time. Dasheen requires at least 7 months of hot weather to reach harvest size. Tubers reach edible size as far north as the North Carolina coastal area. (Where only 6 months of growing weather are hot enough, tubers may be sprouted inside and transferred to the garden when the weather is settled.)

The spot where I grew dasheens last summer is not yet good garden ground, but is much improved from the hard-packed, burned-over parking place it was 3 years ago. By applying gulf grass mulch and spading in kitchen waste, the clay has become workable. The dasheens grew to over 6 feet tall with main bulbs weighing over two pounds. Many smaller bulbs grow out from the parent, and they in turn have bulbs on them.

Tubers weighing from 2 to 5 ounces are used for seed, and may be planted outside two weeks before the last killing frost. If it's possible to plant in the fall, seed gets an earlier start and keeps very well. Otherwise it should be planted in February or March.

Best variety for the U.S. has proved to be the Trinidad dasheen. They are planted 2 to 3 inches deep, 24 to 30 inches apart in rows that are 3½ to 4 feet apart. Weeds should be kept down and soil kept moist until leaves are tall and large enough to shade the ground. Plant several rows in a square or rectangular patch. Then they can shade out all competition from about May until harvest, even if mulch wears thin.

Leaves are cut back and removed first to make harvesting easier. Roots are then lifted in the same manner as sweet potatoes before the first fall frost. The corms are divided, cleaned and stored for use very much like sweet potatoes, but they do not require a curing period. Largest corms should be used first because they do not keep as well as the small ones.

To force shoots for winter use, plant corms near the surface of a sand bed in a warm cellar or greenhouse. Shoots are cut and used as they attain a height of about 6 inches.

The dasheen is a concentrated food containing little moisture and more starch and protein than potatoes or sweet potatoes. It can be cooked much like either of these, although the taste is different—delicious, we think.

No vegetable we grow in Florida is so well adapted to adverse conditions. It can withstand the annual spring droughts best and yet can grow standing in water. It is not relished by insects, rabbits or raiding raccoons due to the acrid quality of the raw plant. By applying organic techniques dasheens are the least work of any vegetable we grow.

—Donna Brunner

EGGPLANT

It Should Be Easy

A vegetable we find easy to raise—despite our being located in short-summer Vermont—is the royal purple eggplant.

We buy our plants from a local nursery, choosing the Black Beauty and Early Long Purple varieties. (The first is the most widely grown, while the second matures slightly earlier and may be better for cool sections.) These are set in the open garden on Memorial Day. Now, Memorial Day in Vermont does not mean the end of cold nights—or frost. But one must get going or there's no harvest.

My husband sets the young plants in a hole which has a deep layer of manure on the bottom covered with soil. Plants are set about 2½ feet apart in 3-foot rows and watered well to firm the roots. Hot-caps and bushel baskets are kept handy, as eggplants like to snuggle down against the cold at night. Jack also lays a heavy blanket of hay around the base of each plant to keep the heat and moisture in.

Sufficient moisture is needed to develop strong flowering buds and well-shaped fruit. The plant itself is attractive and at one time was raised only as an ornamental shrub. Its gray-green foliage forms a striking background for large lavender-shaded flowers. When the blossoms drop, the glossy purple-black fruit appears. Even when the fruit has set and a cool night sneaks up on us, a rush is made to place covers over the eggplants.

Eggplants also do very well in tubs, especially in northern areas where the tub can be placed under the porch or shed during a cold spell. A friend in town has tubs of eggplants on her patio, very decorative and handy for broiling with lamb on skewers on the outdoor grill.

Almost any mulching material is good for eggplants except one—tobacco stems. Here in the Connecticut valley, the stems of tobacco plants are often shredded and used for garden mulch, but remember tobacco is detrimental to eggplants. Hay or straw is what my husband swears by. Mulching is most important, for the plant is a low shrub and the fruit heavy, pulling the stem downward and bringing the fruit to rest on the ground or mulch. The time to apply mulching material is when the soil has warmed thoroughly in late spring or early summer, not sooner when it may delay heating and slow maturing of the plants. Fruit is ready to pick when it has a high gloss. Keep the ripe fruit picked and the plant will yield until frost and later if protected.

—Emma Bailey

Eggplants with Banana Peels!

All the garden books seem to agree that eggplants are a fussy, cranky vegetable, so for years I hesitated to try them. But I find I can provide many of the right conditions—moisture, a soil with much lime, and hot weather. I tried two plants of Early Beauty hybrid last year and was very pleased because 10 or 11 fruits was not bad for a first attempt from just a couple of plants. Since so many of its other requirements were met, eggplants took my heavy soil in stride.

I had heard that eggplants had to have much humus and that too much manure would make the plants soft. I had no compost on hand—so, two or three months before setting the plants out, I began to bury banana peels in the exact spots where I expected to plant my two eggplants. I did this again and again. When the plants were finally ready for the bed, their compost was already buried in place!

As I dug holes for the plants, I buried a handful of bone meal for each one, and gave them both a little liquid manure to give them a start. The eggplants found no fault with their banana peel compost, and soon were sturdy plants with pretty purple stems, numerous lavender blooms, and began producing nice purple fruit. Being from India, they liked my warm climate, and I found them easier to raise than tomatoes, for the latter often drop their flowers when it is very hot.

—Phyllis Holloway

Stretching the Growing Season

There are ways of stretching the growing season for many fruits and vegetables which enable us to grow a wider variety in our gardens each year. Eggplant is one vegetable that must have its growing season lengthened in the colder regions of the country. If the seeds are started indoors early enough in spring, eggplant can be grown with good success throughout the northern part of the United States regardless of the length of the growing season.

Back in Illinois, my father was a great one for growing eggplant and we all loved it. Later, when I married and moved East, I missed the organically grown eggplant. But I was unable to grow my own, and those I bought at the store were spongy and tasteless.

In Illinois the weather warms up early enough in spring so that all my father had to do was sow the seed right in the cold frame. But, up here in the East, the ground is often still frozen during the first week of April.

My father decided there were two things I must do in order to have any success with eggplant in Massachusetts. The first was to soak the seeds before planting to hasten germination, and second, to start them indoors at least 8 to 9 weeks before transplanting time.

He was right!

—Betty Brinhart

ENDIVE

From Spring Till Snow!

You can enjoy endive from early summer until the snow flies.

If I had to choose between a spring and fall planting, I would definitely grow the spring one. Plants are easier to start then, and growing conditions are such that stimulate rapid growth, which gives better flavor and more tender leaves.

There are 3 points one must consider if maximum success is to be achieved with endive. First off, if growing a spring crop, plant it as early as the ground can be worked. In our garden, we plant our lettuce and endive at the same time. The seeds of both of these salad plants seem to germinate best while the weather is still cool.

Secondly, if your soil is sandy, or tends to dry out too quickly, turn in plenty of compost or other organic matter well in advance of planting

69

Endive can be blanched when covered with boards, mats or baskets that shut light out, thus making them tender and sweet.

time—preferably in the fall—and provide irrigation in some form so that the soil is slightly moist at all times. Endive grows rapidly under favorable conditions, and this is very important to its texture and flavor. If its growth is retarded for any reason, especially because of a lack of moisture, the leaves become tough and very pungent, thus ruining your entire planting. Once growth begins, it must continue along on an even keel until maturity if one expects to enjoy endive at its best.

Manures and compost are about the best fertilizers we have yet found for endive. We prepare a special compost pile in late fall to which lime, rock phosphate, and plenty of cow manure are added. In early spring, after we work the plot for head lettuce and endive, we fork a liberal amount of this compost into each row.

Before planting the fine seed, many gardeners mix it with sand or sifted compost. This enables them to plant the seeds more thinly in the furrows. This is not necessary, however, for the tender, young plants can be thinned and used in salads as soon as they are 4 inches high.

In recent years I have been using an entirely different method of planting endive, and I find it gives me excellent results. Instead of dropping the seeds in a continuous furrow made with the corner of a hoe, I plant a pinch of seed in well-fertilized holes every 12 inches along the string in rows 14 inches apart. This not only saves seed, but simplifies weeding while the plants are growing up.

When the young plants are two inches high, I thin each group to stand only 3 healthy plants apiece. A week later, I thin them down to just two plants. When plants are 6 inches high, I thin the rows to stand one plant every 12 inches. Gradual thinning, in this manner, prevents loss of plants to insects and cutworms.

When plants reach 8 inches and begin to form rosettes, I like to loosen the soil well in the aisles with a grub hoe, sprinkle compost in among the plants until the ground is covered, then mulch the entire bed with 4 inches of dried grass clippings, old hay or straw.

Rainfall is usually sufficient in spring to keep our endive growing rapidly. However, if the soil does dry out beneath the mulch, we turn on an overhead sprinkling system and let it run for half an hour, or until the ground is soaked down to 6 inches.

If you are growing endive for the first time, don't make the mistake of waiting until most of your plants have matured before you begin using them in salads. I start using mine as soon as the plants are as large as my fists. The central rosette has not formed as yet, but at this stage the outer leaves are as tender as the inner ones—and just as good. If you wait until all have matured before using them, you may lose a large part of them to rust and rot before you can reap full benefit from your planting. If using endive while still young and tender, blanching is not necessary.

There are several ways in which you can successfully blanch endive. However, blanching is not necessary. Health authorities agree that un-blanched endive, celery, etc., contain a great deal more of the health-giving vitamins than blanched products. But, if you prefer endive with a less pungent flavor and a lighter color, blanch it by all means.

Blanching may be achieved by placing two 12-inch boards on edge along either side of the row so that their upper edges meet in tent fashion to exclude light. Straw mats, overturned baskets or crates may also be used as long as they shut out light. We use overturned fruit baskets covered with several thicknesses of burlap. Clay flower pots may also be used if the small hole in the bottom is corked up. In two or three weeks, the rosette is nicely blanched, and ready for the salad bowl.

—Elizabeth Matuey

FENNEL

You'll Favor Fennel

Fennel can be started in May or June, or a month later for fall use. Where summers are hot and dry, the later planting will do better. Like Chinese cabbage, fennel tends to bolt in hot weather. July 15 (when I start Chinese cabbage) was a good planting date for fennel, here in Massachusetts. Let fennel mature in cool, late September. Depending on variety, it's ready to eat in 60 to 110 days. Good stalks are ready in about 90 days.

Since fennel forms large roots, like others in the family (carrots, parsley, parsnips, etc.), it needs loose, friable soil, free of obstructions. It needs rich soil, too. I used a liberal layer of compost in the bottom of the furrow. Sow seed in rows prepared with two inches or more of compost or aged manure, in drills one-half inch deep and about 18 inches apart.

There was a surplus of seedlings. Some I left in the seedbed, properly thinned to 6 to 8 inches apart. Others remained there too, but unthinned, in a thick cluster. These, of course, never developed. I used them gradually in salads, chopped them up, roots, stalks, foliage and all.

And then, though the directions on the seed packet did not recommend transplanting, I set out about a couple of dozen seedlings between two rows of onions almost ready to be harvested —an example of intercropping. The fennel got shade while it was getting established; and later, when the onions had been pulled, it had room to develop. Note: Because of the long taproot, care must be taken in transplanting fennel. Nevertheless, the transplanted specimens soon caught up with the ones that had stayed in the row and did just as well thereafter.

For crisp stalks, fennel needs plenty of moisture. I could have mulched more and will this year.

Florence fennel makes an attractive, unusual-looking plant in the garden. It stands erect, two and one-half feet tall when mature, with spreading branches, feathery-tipped. A broad, bulbous base rests on the surface. Mound soil about this base if you want to blanch it.

All through fall we used the fennel —how, I'll tell later. Cool weather held the mature plants in good condition. On December 10, with the temperature at about 28 degrees, I pulled up the last plant. The foliage had been touched but not the stalks. Fennel is hardier than lettuce or even cabbage—about the same degree of hardiness as Chinese cabbage. If left in the ground and mulched, the roots would probably survive a cold winter, but there's no point to it unless one wanted to harvest seeds next season. Florence fennel is treated as an annual in most parts of the country, although it can be grown as a perennial in the warmer sections.

Unless a family already buys fennel regularly and likes it a lot, two dozen plants grown to maturity satisfy most needs. Thinnings and young plants should be utilized, too. Remember: (1) Grow fennel more or less as you do Chinese cabbage (though they're not related). Treat it as a fall vegetable. (2) Use fennel, raw or cooked, as you do celery.

—Ruth Tirrell

FETTICUS

For Midsummer Vegetable Planting—Try Fetticus

When the first hot blasts of 90-degree weather cause lettuce leaves to wilt and turn brown, and the head lettuce varieties to shoot up a seed stalk, fetticus is still fresh, crisp and vigorous as ever, and remains so until midsummer when it matures and runs to seed. If any surplus plants reach this stage, we simply dig them up for the compost heap.

All seeds are difficult to germinate in the blistering hot and dry days of midsummer, but we manage it with fetticus in this way: Our garden rows run north and south, and so the east side of the sweet corn rows have afternoon shade from noon on—which provides just the right protection and conditions to encourage growing fetticus. With the first cool days of September, we plant fetticus for the fall garden and find that again it is as hardy as turnips and supplies a delicious addition to salads and greens until killed by a hard freeze.

Here, in brief summary, is how we grow this versatile and desirable vegetable: With the first soil preparation in early spring, we plant the first row of fetticus, covering seed a quarter of an inch and pressing it firmly into the seedbed. A first thinning is made when seedlings are about two inches high and even then the seedlings are excellent for greens and salads. From this time on, successive harvesting is arranged to leave the plants for final growth spaced about a foot apart.

When the outer leaves are about 10 inches long, the plant is loosely gathered together and tied with a soft cord or narrow strip of cloth so that the center will blanch or bleach into a head of near-white, narrow-bladed leaves very closely resembling Romaine lettuce, or the European salad delicacy known as witloof chicory. Plants are blanched and ready for use in 3 to 4 days, but will remain in good condition for several weeks. We try to make several successive plantings at two-week intervals through the spring months. Final fall planting is made to allow about 6 weeks' growth before killing frost. But fetticus will also make excellent growth in a clay pot on a sunny window, and provides attractive greenery as well as a supply of leaves for garnish and salads.

—Dale Hilden

73

GARLIC

Plant It in the Fall!

I had every intention of planting my garlic sets early that spring some 5 years ago. But with so much to do in April, it completely skipped my mind. It wasn't until I was going over the empty seed packs in my garden tool box in early June that I discovered the unplanted sets. What was I to do now? My husband would never forgive me — fresh, organically-grown garlic adds that certain something to his cooking.

Even though I knew it was too late to plant garlic, I did so anyway. They say things happen for the better, but you couldn't convince me of that when I was planting those shriveled but still sound, garlic sets that June.

The growth, as I expected, was poor. Thin, weak foliage died back prematurely—long before any sizable bulbs formed. Because the harvest was nil, I decided to forget about that crop, and let it decompose in the soil.

But the tiny cloves that did form had plans of their own. They lay dormant through the summer under the thick mulch of meadow hay. Around the middle of September, the little bulbs began to stir. One morning I was surprised to see a neat row of green garlic shoots pushing up through the brown mulch. They grew well until the ground froze, then disappeared.

That winter I marked the crop off as a complete loss, and ordered new sets in early spring. I could just as well have saved my money, however, for that forgotten garlic had not given up by any means. As soon as the spring weather turned mild, up sprang those shoots again. They grew fast, and I judged such rapid growth could only mean that they would soon bolt into seed. To be on the safe side, I planted the new sets.

I mulched both plantings, but paid more attention to the new ones. By harvest time in early August, the stalks of last year's garlic were as thick as my little finger, and the bulbs so large they were pushing up out of the soil. Those planted in spring were only moderate in growth, although they received better care.

The foliage began to yellow on both crops about the same time. I was pleased with the results of the spring-planted batch, which I harvested first. But, as I pulled up the plants of last year's planting, I could only stare in amazement. The plump bulbs were twice the size of those I had harvested from the spring-planted row, and bigger than any we had ever produced before.

Ever since that experience, I have been planting our garlic in early September — and harvesting excellent crops every August.

One of the easiest of garden plants to grow, garlic does well in any soil suitable for onions. Soil should be worked deeply, and fertilized well with aged manure, compost or any rich organic fertilizer.

—Betty Brinhart

Green Garlic Shoots— an Herb Garden Treat

Although light-green garlic shoots will sometimes appear within 4 days after you plant the separated cloves, they usually seem to take forever to come up through the soil. They're well worth waiting for, however, for green garlic—more subtle and delicate than the cloves—enlivens the taste and aroma of an endless variety of salads, soups, stews, fish sauces, meat dishes and chutneys.

Garlic can be planted among the perennials, but we prepare a small rectangular bed at the south or east end of the vegetable garden for ours. A slightly raised bed is best, so that the garlic will not be overgrown by larger fast-growing vegetables. Where the soil isn't quite fertile, we add old compost plus some ground rock mineral fertilizers to it before planting.

After we level the bed with a rake, we stab holes 6 inches apart slightly larger than the width of a garlic clove, and deep enough so the top of the clove lies an inch or so below the surface. Gentle pressure with the metal rake head or a wide board firms the soil around the cloves after they have been dropped in, *points up.*

Press garlic cloves when planting firmly into the soil so they stand up in holes.

We cut and use green garlic daily, often twice a day, so our plants never flower; they are too busy growing new leaves. Left alone, they would produce lavender blossoms, and the planted clove would become a bulb of tight-packed garlic cloves. (Bulblets grow in the flowerheads, also.)

Garlic repels Japanese beetles and discourages aphids; many organic gardeners plant the cloves close to susceptible plants.

—Devon Reay

Elephant Garlic— Flavor Plus Aroma

In elephant garlic, the most important considerations are the delicate flavor and mild aroma. Those with educated palates, accustomed to the vigor and strength of ordinary garlic, may find the elephant variety something of a weak sister. But for those who have never been able to face garlic in the past, this new and gentler vegetable will prove a welcome taste treat and a very worthwhile addition to the family diet.

All garlic is easy to grow, and the elephant variety is no exception. One thing to keep in mind is that garlic is a cold-weather plant. Frost and even light freezing will not harm it. Indeed, in many parts of the country and in all the southern and mid-south states, garlic should be planted in the fall. In good organic soil the cloves will sprout quickly and make vigorous growth. By mid-May, the plants will be shooting up flower stalks, and these should be promptly cut back and discarded so that maximum development will be confined to the root and bulb.

With warm summer weather, garlic ripens and the plant leaves begin to yellow and dry. This is the signal to lift the bulbs. Allow them to dry thoroughly in the shade for 3 or 4

Elephant garlic is mulched with clippings heavily, and planted in 80-foot-long rows.

The ideal soil for growing elephant garlic is a light loam with an ample supply of humus. However, here on our own place where we have a heavy clay soil, we have overcome its short-comings by the addition of plenty of barnyard manure and proper methods of cultivation, and can produce 8 tons to the acre or better of elephant garlic. To get these high yields the garlic should be planted in 24-inch wide rows, and 8 inches apart in the row. On an acre basis this figures about 30,000 bulbs to the acre. Under a little better than average conditions, 3 bulbs should go 2 pounds.

Precautions: Cut the flower blooms that will develop on top of elephant garlic plants. If allowed to develop and go to seed, it will result in a much smaller bulb of garlic. It does not pay to save the seeds and try to build up a stock from them. A pH of from 6.5 to 7 is satisfactory for elephant garlic.

—Dale Hilden

days, then knock the loose dirt from them—do not wash—and store them in a moderate, dry temperature. The bulbs can be spread out on a rack where they will have good air circulation, or the tops may be left on and braided so they can be hung up in strands for storage.

You wouldn't dare serve common garlic as a vegetable dish, but the mild, sweet, large cloves of elephant garlic are something else again. Try them gently steamed in a double boiler and served with butter sauce or sprinkled with buttered bread crumbs. With common garlic, you merely rub the salad bowl to get a whiff of potent flavor, but with elephant garlic you can slice the mild cloves directly into the salad, as you would onions, or cook and serve them right along with the family pot roast. Elephant garlic is not just a fragrant aroma, but a really mild, wonderful, health-giving vegetable.

Oriental Garlic

Neatly attractive, Oriental garlic can fit into the flower bed to save space in vegetable gardens. Use it as a border, perhaps alternating with chives. Or try single clumps located in a sunny rock garden. Dried flower heads on their stiff stalks are suitable winter bouquets.

As for growing, Oriental garlic plants do well in good garden soil, not over-rich. Divide clumps every other year or so, then reset to form new plants. You can also save seed to sow in spring. My own first plant came one September from a Rhode Island herb garden. As advised, I cut it back when setting out, then mulched over winter —not strictly necessary, but a good practice. New shoots pushed up the next spring, a little later than those of regular garlic, usually the first growth in my vegetable garden.

To gather for table use, cut off outside leaves or blades all the way to the base, instead of cutting off a few inches from the top of the whole clump. This will stimulate constant new tender growth. Sometimes, of course, there may be need to level a whole clump to the ground. It will grow back quickly, though.

—Ruth Tirrell

Garlic and Roses

Garlic among roses brings a number of benefits to the flowering shrub. According to those who have studied how plants exude an essence or essential oil into the air, soil and water in their environment, the two make ideal companions. Many rose growers report that garlic planted with their bushes has prevented insects. It can also be an ally in efforts to banish such rose problems as rust, mildew and blackspot, besides mites, thrips or any of the beetles which seem to favor feeding on roses.

The soil in most rose gardens would be a paradise for garlic, which also loves rich, well-drained land high in humus content. If the soil around your rose bushes is already fairly loose, not much digging, if any, is necessary to place the divided cloves about two inches deep in the ground. If the small bulb seeds from the top are used, about one inch would be enough coverage. Seed can be sown rather thick, especially if you want to use the plants for ground-covering greenery or plan to

use some for salads and in cooking. (I have found that green garlic is delicious both ways.) A tool such as a bulb planter does the job well. Where you want a few set closer to bushes and hesitate to dig because of possible injury to rose roots, just lightly scratch the surface of the moist soil and place bulbs near the bushes, then sprinkle some soil over them.

One important need in getting garlic plants sprouted, growing and well-rooted before cold weather is to provide enough moisture if rainfall is slack. Another aid is to soak the cloves a few hours before planting. We have also left them outside on damp ground in the semi-shade under a grapevine or tree for several days where we kept them very moist.

After the bulb sections swell and perhaps a few roots appear, they may be placed in moist soil with the roots downward. A light mulch in addition to a sprinkling of at least one inch of soil would help to hold moisture where an extremely arid environment exists. But usually the extra sunlight without additional covering will aid earlier and greener development of the tender blades—so it is generally better to mulch after the blades are up a short distance above the ground. We have found that a mulch of leaves, straw, yard trimmings, etc., helps both garlic and any other nearby vegetation by holding moisture and shading the ground when direct, hot sun would otherwise cause rapid evaporation.

—Mayme Bobbitt

GOURD

Drop a Seed and—
Run For Your Life!

In our area of southeast Iowa the advice to gourd growers is "Drop a seed and run for your life!" The reason for that is that a vine from one hill of seeds leaped out to:

1. Cover the garden fence;
2. Ramble over the fence and across the cement hog floor;
3. Camouflage and cover everything in an area 900 feet square!

Only organic fertilizers are used, and we don't have to worry about chemical sprays for insect control—the plants take care of themselves! Sheep manure, house garbage, leaves from the strawberry mulch, plowed-under garden produce, plus corn silks and husks from the sheller all go into the garden reservoir of fertility.

While we're proud of the size of our parsnips, carrots and tomatoes— we feel they speak for the organic way of gardening—we urge you to be sure and plant at least one dipper gourd. And when you do—stand back and get out of its way!

—Ellen R. Fenn

HORSERADISH

It's Attractive and Zesty

More than once we have had a barren area along a west or south foundation much in need of a good foliage plant that would thrive in poor soil or sandy soil very near a much-used foot path. The first plant we tried, horseradish, was such a success that it became our favorite foundation perennial.

Horseradish (*Radicula armoracia*) is a large herb, two to three feet high, with lance-shaped leaves usually delicately scalloped, but occasionally deeply serrated. The tall heads of clustered small flowers are creamy white, the roots pungent. In rich garden loam it spreads quickly, but in poor soil it remains where planted, the lush green foliage filling bare spots beautifully.

78

We were not happy with the appearance of our west foundation till we pulled out bearded iris, hosta and hollyhocks planted by the previous owner, and put in horseradish. The plants have always been healthy and attractive, free of disease and insects.

The white roots have a sharp bite; when they are grated and mixed with vinegar and seasonings the taste is pungent and stimulating. Pickles of excellent flavor can be made by adding a two- or three-inch section of horseradish root, sliced once lengthwise, to small fresh cucumbers fitted in a quart jar, immersed in cider vinegar along with a tablespoon each of sea salt, mustard, and raw sugar. They are ready to eat after a month in a cool, dark place.

—Devon Reay

JERUSALEM ARTICHOKE

America's Oldest Vegetable

Don't underestimate the Jerusalem artichoke. One spring, after enjoying our first taste of that crisp, delicate-flavored vegetable, we could hardly wait to put 6 of the knobby tubers underground in the northwest corner of the garden. Previously we had grown some horseradish there along the fence, but it hadn't surfaced yet. So we set one artichoke tuber practically on top of a horseradish. It didn't bother the artichoke, though—one plant that can compete with horseradish without coming in second.

We planted artichokes in the same corner so that lower-growing vegetables, including sweet corn, wouldn't be shaded by the tall sunflower-family plants. But we needn't have troubled, for even before they sprouted, furry new-green artichoke shoots appeared all over the garden. We had buried fruit and vegetable trimmings —among which were thin artichoke peels—all over the garden to augment the soil's humus content. So whatever we plant, up comes artichokes! If we weren't so enthusiastic about the tuber, probably the oldest native American vegetable, we might have too much of a good thing.

Early in the 17th Century, when French explorers in the St. Lawrence area saw the Indians eating Jerusalem artichokes, they cooked and ate some of the white-fleshed tubers. Finding them palatable, the explorers sent some back to France where they soon became popular and were called "pommes de Canada" (Canadian apples) or "batatas de Canada" (Canadian potatoes). Their cultivation spread to Italy where they were called "girasole articiocco" (sunflower arti-

JERUSALEM ARTICHOKE

Jerusalem artichokes yield heavily—up to 3 or 4 times as much as white potatoes.

September, dozens of blossoms with bright-yellow petals like black-eyed Susans open at the top of each stalk. The plants become so top-heavy that they sometimes topple over in rainy, windy weather, but still lift their crowns 5 or 6 feet high and flower.

About a month after the blossoms fade, when the plants have become dark and dry, you can dig the crisp tubers. In old New England these completely starchless tubers were pic-kled or boiled and served in a cream sauce. They are delicious french-fried, but best raw, especially served in fresh salads.

The longer they remain underground, the better the taste. You can leave them in the ground all winter, just piling pine needles over them. and perhaps covering with overturned wooden boxes before deep snows are expected. Then you can dig what you need during thaws or unseasonably warm winter days. One New England farmer says, "The best time to dig and eat artichokes is early spring, when potatoes start to show their age."

choke) and grown in the famous Farnese gardens. About 1621, the English became interested in the vegetable, but mangled the pronunciation —which accounts for the name Jerusalem.

We plant the tubers two inches deep, more than a foot apart in sandy soil fortified with old compost and loosened with a spading fork. Young plants need weeding only once, for they quickly get bigger and tougher than the weeds. However, they can be mulched with pine needles, grass clippings or other suitable materials.

By the time the plants reach an 8- to 10-foot height, about the last week in

Girasoles have been more appreciated in Europe than here, except as livestock food. In France they have been cultivated and improved for over 300 years. Now, a new popularity is developing for America's oldest vegetable. The best of some 200 varieties is the American artichoke, formerly called the Improved Mammoth French White, a large uniform artichoke with a delectable fruit-nut flavor.

One thing about artichokes, they practically raise themselves. They yield heavy, too—up to 3 or 4 times the harvest of white potatoes.

—Devon Reay

KOHLRABI

Don't Miss Out on Kohlrabi

Kohlrabi grows most vigorously in cool weather and requires plenty of moisture. It should do best in a soil slightly acid to neutral, yet ours thrives in the same soil in which we grow potatoes, lettuce and cucumbers.

As early in spring as the ground can be worked, we loosen the soil with a spading fork, then spread compost two inches deep over 10-inch-wide rows. (In making compost, we believe earthworms and moisture are important factors. If the pile is moist enough, very little turning is needed; the worms work through the material and convert it to humus. We added 1,000 red earthworms to our heap 3 years ago, and they have multiplied ever since. When we start a new compost pile we always include more than a bushel of old compost, which of course is alive with worms of all sizes.)

We spread a little garden soil over the compost, rake it level, and sow the seed very thin over the 10-inch-wide area, the length of the row. If we sow lettuce or parsley in the same row at the same time, it is sown even thinner. The seed is pressed into the earth with the flat head of the metal garden rake so that it is about ½ inch deep. Rows should be about 18 inches apart, and the seedlings thinned to 6 inches apart within a few weeks. To maintain adequate moisture, we mulch the small plants heavily with grass clippings, hay, pulled weeds, or even compost.

We plant Early White Vienna, a light-green, white-fleshed variety; we can cut the bulbs after 55 days. Successive small plantings, two weeks apart, can be made till mid-May; in August seed can be sown for a fall crop. These plantings insure a constant supply of small bulbs—two to two and a half inches—the size at which they are most tender and succulent.

Don't pass up these simple, easily followed, how-to rules:

1. Plant as soon as the ground can be worked in the spring—they are cool-weather growers;

2. Thin to four inches apart and cultivate lightly;

3. Mulch heavy and give plenty of water—keep the soil moderately moist at all times;

4. Grow them fast and pick them early—within 80 days.

A final word. Keep that ground working and improving after you have gathered your kohlrabi. Plant it to snap beans and then turn them under when they begin to form pods. They will release plenty of nitrogen right into the soil as they decay. In this way you will fertilize your garden and get it ready for next year with very little extra effort.

—Dexter Raymond

LEEK

The Poor Man's Asparagus

Leeks are so popular in Wales that they are the Welsh national emblem and are considered the "poor man's asparagus." I dig mine up regularly, all winter long.

One snowy 15-degree day last February, I dug up a dozen thick-stalked leeks for cock-a-leekie, a Scottish soup in which chicken and leeks are the main ingredients, with potatoes or rice or barley added to thicken. The leeks thawed out crisp and firm, and I peeled off only the outer skin—many closely-folded layers make up the white stalk and its slightly rounded base. Even the few inches of blue-green leaves at the top—which had stuck up above the deep mulch—could be used. By contrast, Chinese cabbage at that temperature thawed out to a mush and couldn't be used.

Leeks take a long time to grow—4 to 5 months. The root system of a mature leek—dozens of thick threads—is evidence of long, slow growth. I sow seed in early spring for fall leeks and a little later for use in winter. Where winter is mild—say in Virginia—half-grown leeks, started in early fall, might be preferred for wintering over, since they start growing again in late February.

Leeks benefit from transplanting. I usually set the seedlings—which look like tiny blades of grass—into the surface of the soil, then later bank dirt about the growing stems to blanch them. A reader from Enumclaw, Washington, whose father raised leeks by the acre for the London market, advises: "Plant in a shallow trench, then fill in as they grow—it is much better than hilling."

Full-grown leeks should stand about 6 inches apart, so I set the seedlings at that distance and don't count on thinnings. Leeks may be served raw—so can Egyptian onions—but when cooked, they surpass most vegetables for savory appeal, reaching peak perfection when fully mature.

Popular, reliable varieties include Large American Flag — sometimes called Broad London — which is hardy, vigorous and a good winter plant. Elephant leek is recommended by Harris for fall use and I like Conqueror, a slender-stalked, hardy grower, for winter eating. An older, once-popular strain that no longer seems available, the St. George was planted for fall and winter, although it was not classed as especially hardy.

—Ruth Tirrell

LETTUCE

Harvest 10 Months A Year

You can grow lettuce over 300 days of the year. The only extra requirement for such a steady supply is a cold frame exclusively devoted to lettuce-growing. This can be bought ready-made or it can be made—except for the sash—by anyone who can wield a saw and hammer.

The broad plan of this program is to rely on the cold frame for lettuce during the early spring and late fall, when freezing prevents growth in the garden; and during midsummer, when garden conditions are too hot for lettuce. The garden is used during spring and fall, the best seasons for ordinary cultivation of lettuce.

Things to remember at all times are:

1. Lettuce is hardy, but will not withstand freezing; it tends to go to seed prematurely in dry, hot weather.

2. Lettuce must be kept growing rapidly if it is to be lush and tender and if the heading varieties are to head. As lettuce is a leaf crop, it needs plenty of nitrogen-rich organic materials.

3. Leaf lettuce will stand a little crowding; romaine even less; head lettuce none at all. It is the nature of head lettuce to spread, and it is almost the invariable rule that if head lettuce, in season and well-nourished and watered, fails to head, the cause is crowding.

4. Cultivation should be frequent to hold down weeds and keep the soil surface loose, but light, because lettuce roots are near the surface. A mulch is particularly helpful.

The cold-frame soil for lettuce cannot be too rich in organic fertilizers. Spread at least two inches of compost or well-rotted manure (chicken, preferably) on the top and mix it well with the soil. If dried manure is used, the quantity can be a third smaller. Since lettuce is a shallow feeder, it isn't necessary to work the fertilizer in more than 4 or 5 inches deep.

If running water isn't handy, a barrel can be placed where it will catch the runoff from a roof. Throw an occasional forkful of manure into the barrel or a few trowelfuls of dried manure. If you don't use a barrel, make up an occasional batch of liquid manure in a pail.

A watering can without the sprinkler cap on the spout is best for cold-frame use. The soil must be soaked well, not sprinkled, and water should be kept off the leaves as much as possible. Besides the liquid manure, also spread an organic fertilizer twice a season; say, after the spring crop is harvested and after the summer crop. A pound or two each time should be enough if the soil is well enriched organically.

Now for an approximate timetable of what to do. The following schedule is only approximate for average temperature conditions around New York.

It is subject to variation not only because of climatic conditions in other temperate zones, but from the vagaries of the weather, the varieties planted, and the nature of the gardener. In short, everyone has to work the details out for himself, and vary the program in accordance with growing conditions.

February

Get the cold frame started. The ground will begin to thaw if the sash is kept on the frame, probably by the middle of the month. The warm-up will be hastened if the mats are put over the sash at night. When the ground can be worked, add compost or manure, if this wasn't done in the fall; turn it over with a fork, pulverize it well with a rake, firm it down by treading over it on a plank, rake very lightly again, and all will be ready for the seedlings, which should go in around the first of March.

Grow the seedlings in a flat placed in a very sunny southern window. Fill the flat nearly to the top with soil from the cold frame, firmed down well. If your soil is of very heavy clay texture, add up to 25 percent of sand or shredded peat moss because seedlings in heavy soils, under confined conditions, are subject to "damping-off." This is a fungus complaint which destroys young seedlings.

With the soil rather moist but not soggy, mark off rows 1½ inches aprt, and make quarter-inch-deep drills for the seeds. The time is about 3 weeks before you expect to set out the seedlings in the cold frame. Sow the seeds rather thinly, but without gaps in the drill, and cover with ¼ inch of fine soil or (preferably) sand. Sprinkle lightly, then cover the flat with glass, wood or paper to hold the moisture until the seeds germinate, which

should be in a very few days. Then the cover is removed.

Keep the seedlings watered and begin to thin them as soon as practicable, as crowded seedlings quickly grow spindly indoors. You are aiming at an ultimate stand of seedlings just 1½ inches apart, in the rows as well as between them.

A week before transplanting time, run a knife through the soil down to the bottom of the flat, making equidistant lines each way so that each seedling can be lifted out with most soil intact, roots unharmed. Almost a week before setting out, begin to harden the seedlings by reducing sharply the amount of water they have been getting, so soil becomes almost dry. Wet the soil again before transplanting.

March

Remembering our observations about crowding, the plotting out of the cold frame for the seedlings of the various types of lettuce deserves a little care. A foot apart and between the rows is about the average distance to figure on. Space leaf and romaine lettuce about 9-10 inches; head lettuce about 14-15 inches. An average of a foot apart means the 6' x 6' cold frame will take care of 36 maturing lettuce plants at a time. Set out about twice that number, an average of 6 inches apart, discarding the weakest seedlings in the flat.

Assuming a setting-out time of March 1, by March 15 you should be able to start eating lettuce, thinning out the alternate plants of the leaf variety you have planted. When the soil in the garden can be worked, set out the alternate plants of the crisphead and romaine varieties from the cold frame, and begin sowing seeds in the garden for succession crops, two weeks apart.

During cold weather, the cold frame requires close attention. On sunny days it needs ventilation—move the sash down to make an opening of a few inches to two feet at the high part of the frame. Any gathering of moisture on the glass is a sign that ventilation is needed. Well before sundown, close the frame to retain the heat, and unless the weather is quite mild, cover at night with the mats. Watch the condition of the soil, and water at any sign of dryness. The soil in a cold frame uses an unbelievable amount of moisture; as the weather warms, the need for water increases constantly.

April-May

These months you may leave the cold frame uncovered, except when the temperature gets down near freezing. Continue to sow succession rows in the garden; thin, cultivate and water if rain doesn't fall freely once a week. Harvest and eat your lettuce as it matures in cold frame and garden.

June

By some time this month, the cold frame will have been cleared of mature lettuce. Just before real warm weather sets in, transplant young seedlings from garden to cold frame, leaving room to sow seeds in the frame for late-summer heads.

July-August

Hot-weather care of the cold frame involves use of the slat covers already referred to. These go on when the sun shines; they are taken off at night and kept off in cloudy or rainy weather. This, plus constant watering of the soil to keep the surface moist—as often as every other day—is the secret of tempering the sun and keeping cool enough conditions for satisfactory growth of the lettuce.

By late August, you may set out in the garden some of the seedlings developed in the cold frame and start again seeding in the garden, spacing small seedings two weeks apart.

September-October

Harvest the garden and cold frame and, before the first frost, fill the frame with young seedlings from garden sowings made in late September or early October.

November-December

Your problem at this time is the same as early in March. Constant vigilance is needed to carry the plants over the first freezes. If all goes well, you can continue to harvest through a good part of December.

Storage

If you have arranged things so that a dozen or two of good, firm crispheads are still in the frame as a heavy December freeze approaches, pull them up for storage. Trim loose leaves and cut off the roots. If you have a cooling chamber, such as in a "walk-in" freezer with a temperature close to freezing, you are particularly well equipped. If not, a cold cellar kept very moist will do quite well.

With luck and good management, you will have lettuce for two or three weeks of storage. This will carry you into January, almost time to start the window flat again!

—John Gourlie

Lettuce Under Netting . . .

If you want crisp, green, flavorful lettuce all summer long—grow it under netting.

Stretched across old slats, odds and ends of old fencing and poles, netting cuts the heat of the sun by as much as 35 to 45 percent. If you use cheesecloth, double it over. To make sure my netting will do the job, I measure the light *above and then below* the netting with a photographic light meter. It tells me how much light I am cutting out and guides me when I am rigging my nets.

When its readings tell me I have cut the direct burning sunlight down to the above limits, I know my netting will protect the lettuce and keep it from wilting and drying out. Of course the soil must be kept moist and fertile all the time it is growing successive crops of lettuce during the hot summer days. But you will find that your nets positively reduce soil evaporation and plant transpiration.

Lettuce in shaded beds transplants easily, so I frequently sow 3 feet thickly of a favored variety in a single row. I permit this to grow densely to a height of two inches and then transplant directly to a final growing site. Here I do not set them too closely, allowing 5 to 6 inches between plants.

—Gordon L'Allemand

. . . and by the Can

Lettuce was something I couldn't have. It came up—then disappeared. I'm not sure if it was pigeons, slugs or grasshoppers that were responsible. Then I tried planting it in tin cans. Now I have more lettuce than I can use. The lettuce grows a long stem—comes up above the can—which is set at least two inches in the soil. Have been using different size cans. Also found this very beneficial during our 3-month drought from December to March 23rd, as I could water each plant directly—and it soaked down.

Last year I tried growing tomatoes in large cans. Have had very good results. By having them in cans I'm able to keep them in partial shade, which serves two purposes—protec-tion from very hot sun and from too much rain and eventual blight. The growing season here isn't cooperative. We have months of no rain—then again months of daily pouring rain. Gardening is a real challenge—starting with a pure clay soil, too. But I've used compost and mulch—now the soil is much improved.

—Mrs. Lena Forbes

Blood Meal Saves the Lettuce

I had finally changed the jungle of briars, ordinary weeds, and extraordinary weeds like Canada thistle (locally called "candy thistle") into a more or less plantable place. The past winter I covered the cut-over space with newspapers weighted down with a mulch of leaves, and when spring came around, without further cultivation I just made rows with a hoe and planted potatoes and fresh Salad Bowl lettuce seed.

Was surprised and pleased that so much came up, and was especially pleased with the lettuce. Proud of it, I took callers to see it one afternoon and saw that it was all eaten off. Imagine my surprise and feelings! Woodchucks, of course.

I got a neighbor proud of his trapping ability to bring traps, and he caught 6 of them, while I ignored the lettuce row which began to grow again from the undamaged roots.

Discouraged, I paid no attention to that part of the garden until I noticed perky rabbits taking an interest in it. Then I sprinkled blood meal along the rows—and the rabbits lost interest. The lettuce grew and grew, so that I had plenty for my own use and lots to share. Blood meal is often the gardener's best friend.

—Edna Hindle

Protection with a Plus from Clothes Hangers and Wire

To protect young lettuce, peas and beans from rabbits in the spring, I make row-long, reinforced U-shaped tents by weaving long pieces of straightened clothes hangers through different widths of chicken wire. It is very noticeable how well these tender young plants do under the metal tents. Later on, I move the larger tents over to the ripening ever-bearing strawberries to keep the birds out.

It is my most earnest belief that this controlled feeding and watering, *aided by the atmospheric electricity conducted by the metal tents, produces bigger plants, more abundant, taller foliage, and berries of distinctly superior size, color and flavor.* Nearby, my untented June-bearing berries do not produce this quality plant or fruit although they receive the same feeding, watering and care.

After freezing weather has stopped all growth and bearing, the strawberry rows are mulched heavily with well-shaken-out straw, and the reinforcing wire is removed from the chicken-wire tents which are laid back over the mulched rows to hold the straw in place over the winter.

Combined, my stock wire and iron pipes do a good job of supporting many kinds of vining vegetable and fruits, keeping them well up in the air. I believe the extra height permits them to attract more electricity, while the improved air circulation reduces fungi attacks. It is not surprising that they all produce larger, more colorful and flavorful fruits and vegetables with much less spoilage than those which are permitted to sprawl out over the ground.

—Edward P. Morris

Beating the Heat

Even the drought these past years has had its good points. In gardening for the most flavorful, succulent vegetables (the only kind worth growing), there is no compromise.

Head lettuce is one of my favorites —and I am not satisfied with just a few heads squeezed into spring before the heat arrives. I want it all summer. Until the drought hit this normal eastern Ohio climate of adequate rainfall and no prolonged heat, there were no problems.

The incidental problem of any drought is low humidity in combination with prolonged sunshine. Transpiration is so rapid then that avoiding wilted-in-the-garden lettuce is almost impossible. But I persisted, and came up with one solution. The simple wire frame and cheesecloth cover I use is easy to ready and handle, can be set out or removed quickly and stored from season to season. Wooden slats in a lath-type shading arrangement also provide an effective hot-weather technique for lettuce. Some gardeners capitalize on shade protection from fences, nearby shrubs and trees, or from purposely interplanted taller vegetables and ornamentals.

Even without a drought, this is a way of growing any number of cool-season vegetables and flowers in normally hot climates. It is well worth the little effort and minor expense.

More Lettuce From A Small Patch

You can have a big bowlful of lettuce every day from a patch about the size of a card table. I plant a little more than that, but then we have it twice a day and keep giving it to neighbors.

Outside of working a layer of compost into the soil surface, there are just two things I do that guarantee a steady supply of crisp, crunchy leaves of lettuce. One is the manner of planting, the other the way of harvesting.

Long ago I discovered that there was no reason lettuce had to be planted in rows, unless one were raising it on a large scale commercially. I sow the seed rather sparingly over the entire patch. As soon as it comes up I thin it so that all plants are 1½ to 2 inches apart. This gives each plant an adequate supply of nourishment and plenty of sunlight. For rapid and sturdy growth keep the soil damp.

The most wasteful way to harvest lettuce is to pull out the entire plant. A second method, which is a vast improvement but still wastes a lot of good lettuce, is that of cutting off the tops of the plants. They will grow a second crop, but it takes quite a while.

The way to have a steady flow of lettuce to the table is to snap off the outside leaves. This way the smaller inside leaves are not damaged, and they quickly develop to harvesting size.

My two favorite brands over the past 30 years have been Prizehead and Bibb. These two give just the right variety for daily use.

—L. T. Servais

By the Panful in the Off Season

Fall-grown lettuce can be transferred from the organic garden into two- and three-inch flower pots, and its growing life extended far into the winter. I had wonderful success by going about it in this way:

The fall-grown lettuce shoots were pulled, roots and all, and the tender leaf portion was cut off down to the last two shoots. The shoot, with the root, was then set into the pot and set into the pan. In a few days these shoots started to develop and as they grew larger still other shoots showed up. As they became large enough to use, they were picked off and used in sandwiches as relish and in other ways.

The lettuce was watered daily by pouring water into the pan and allowing the water to enter the pots from the bottom. Fast growth of tender leaves is assured if good organic soil, derived from a compost base, is used.

We used lettuce from the potted plants until after Christmas last year. We like the idea so well that we are repeating the lettuce project again this fall. We gave the lettuce all the outdoor living possible, but when hard freezes started we kept the pan under lights in the basement and a plastic cover over the pots to increase the air moisture around the lettuce leaves. Otherwise dry edges may result.

—Emil G. Glaser

Cos Lettuce—Tender All the Way Up

When we moved from the country last fall to a house in town, we had to adjust to a garden space 32 feet long and 32 feet wide. That's why we decided to concentrate on salad plants, with special attention to that king of them all—lettuce.

From the 6 types of lettuce (crisphead, butterhead, cos, leaf, semi-loose heads, and stem) we chose cos because it stands the summer heat in town. It looks attractive, too, in garden rows with its long, fairly loose leaves. Besides, it forms a head without even being tied. And best of all, it's tender all the way up. In terms of space and care required we felt cos would be most rewarding—as it left time and room for tomatoes, radishes, cucumbers and cabbages.

Several reliable varieties of cos lettuce include Balloon, which is a light-green with big outer leaves enclosing the heart and giving an effect rather suggestive of a balloon; Hick's Hardy Winter White, which can be sown in autumn in moderate climates, to heart up in spring (although it does tend to spread); and the variety we chose—Lobjoit's Green Cos— which is probably the best known and most popular. It is large, medium- to dark-green, and does not require tying.

Before planting, we chilled the cos seed on blocks of ice to prevent it from going into dormancy when planted. Cos is a cool-weather crop and should be planted just as early as ground can be dug and worked. When the hard freezes were over in our part of Ontario—which is usually in April—we sowed the seed directly outdoors. We tried not to sow too thickly as this would only mean extra work later on when thinning, since overcrowded lettuce cannot develop properly. Another precaution: covering small seeds too deep results in poor germination, especially in wet weather; a soil protection of one-quarter inch is sufficient. We cultivated just often enough around the plants to maintain a loose, weed-free surface.

There were wind-drying days in early spring, and we watered frequently with a fine mist-like spray so that the seeds would not be washed out. A sandy soil such as ours fills up with water faster than one that is mostly clay. But the reverse is also true, as sandy soil will dry out faster. Once the seeds germinated and little plants pushed up to sunlight, we tried to get as much water into the ground as possible without washing away any soil in the runoff. Evening, or when the sun wasn't blazing down, was our watering time.

When our plants reached about 2 or 3 inches tall, we thinned to 2 inches apart to prevent crowding. As they grew higher, we thinned them to 4 inches apart. A final thinning of 8-inch spacing came at the period when the cos began to head up.

The time to harvest cos is when it become cylindric, with the head almost closed. We never picked ours right after a rain because plants fill with water and break easily; also putting it in the refrigerator with too much water on encourages rot.

—Lorie Porter

For Lettuce—Use Your "Head"

Despite the superior flavor and "meatier" crispness of the so-called cabbagehead, or simply "head" lettuce, many gardeners shy from its cultivation in the mistaken belief that it is difficult to grow. The problem, it is said, is the failure to form heads. I wouldn't know. In the 10 years I've grown this type of lettuce, I have never encountered the trouble. Here's what to do:

Sow seeds for succession rather than all at once. How long the succession is the question mark in cultivating this type of lettuce. In climates moist and cool enough for the cultivation of a spring crop, gamble with a small garden area reserved for this special vegetable. As during the past two summers, August may turn out surprisingly cool. And if it doesn't, there are ways of beating the normal heat.

Plantings that will be brought to head during the hottest part of sum-

LETTUCE

mer should be located with some shade. If no natural shade exists, construct a simple low framework covered with laths spaced an inch or two apart, or with several thicknesses of cheesecloth—either of which tempers the ill effects of the hot sum without materially shutting out the required light.

In any event, a late sowing should be timed to bring the plants to head during the cooler days and nights of autumn. I usually make my last sowing about the first of August, so that plants will not be ready to form heads before the coolness of mid-September is reasonably assured. Using a cold frame for still later sowings, you own crisp head lettuce can grace the Thanksgiving or even Christmas table.

Most important for any leafy crop is a soil especially rich in nitrogen. Late in autumn I mix some dried cow manure into several garden rows where the first lettuce will be grown. The following spring, when the seedlings have begun to fan out into leafy rosettes, I spread a circle of cottonseed meal around each plant and work it into the plot.

In soil already rich in nutrients, I consider having adequate water more important than additional fertilizer. Enough moisture means rapid plant growth—which produces a crisp head of lettuce. To improve the soil's moisture-holding ability, I incorporate

a good amount of compost as it is prepared for a succession crop after lettuce is harvested. This assures a newly enriched growing medium for the succession crop and maintains the humus content of the soil from year to year.

After the ground has begun to lose the chill of early spring (about the first of May), I apply a thin straw mulch around and tight up underneath the lettuce leaves. This does 3 things: holds soil moisture; keeps the large leaves off damp soil to prevent rot; and maintains the cool root-run that many plants (especially cool-season vegetables) require for best performance. As warmer weather arrives in June, the mulch is gradually deepened to 3 or 4 inches.

Avoid crowding plants. Young plants should be spaced at least 15 inches apart, and 20 or more is not too much in soil rich enough for a bumper crop.

Most varieties of head lettuce found in today's catalogs for home gardeners resist tipburn and are reasonably sure headers during warm weather. My personal preferences are Great Lakes and its slightly smaller and earlier strain, Pennlake. Other good heat-resistant varieties are Iceberg, New York No. 12, and Imperial No. 44. And from your own patch, they all produce flavorful, tender crispness you never taste out of a supermarket.

—Richard D. Roe

LUFFA

Patience and Care Bring Bountiful Results

Because it is cold-sensitive, luffa should be seed-sown outdoors only when the ground is thoroughly warm. Be patient if it seems slow in sprouting and, if your growing season is really short, start the seeds indoors individually in pots, adding some peat moss to your potting mixture.

Transplant outdoors when all danger of frost is past, picking a balmy, cloudy day for the operation. Plant the seedlings 2 to 4 feet apart in rich soil where they will get plenty of sunshine and where you—eventually—will get plenty of good organic sponges.

For best results, the plants should be supported by a trellis or wire fence, so the gourds won't touch the ground. Cucumber-shaped, about 12 to 20 inches long, the luffa is grown much like the cucumber, attaining a height of 10 to 15 feet, and producing pretty, light-green foliage and whitish blooms. Be sure to water deeply all through the season for vigorous, healthy growth.

The young leaves may be eaten any time, while the young fruit is used until it gets as large as cucumbers. When properly grown and given lots of sun, the fruit matures before frost, attaining a full two feet. If you want fine-textured white sponges, be sure to harvest the gourds as soon as they turn yellow, leaving enough stems so you can hang them up easily to dry.

The drying process should be gradual and is best done outdoors in a shady place. Peel the gourds when they are dry but still greenish, and push out the black seeds. Next, wash out the fibrous centers and remove all loose tissue. After drying, cut in half or into smaller pieces, depending on how you want to use them. Some people like them the full length of the gourd.

As many as 25 fruits may grow on a single vine, each weighing 5 pounds or more. Better fruit will be produced by pruning off all first flowers and poorly shaped gourds. As noted, the fruits benefit by being kept off the ground, and should be supported by flat stones or some sort of trellis or fence.

Seeds to start you off are available from seed and gourd dealers. After you raise your first crop, you can save your own supply, drying them in the sun and storing against next year's plantings. Seeds thus treated retain fertility for 4 or more years.

—Robert J. Wyndham

MUSHROOM

Mushrooms in the Cellar . . .

We live in that section of Arkansas known as "Tornado Alley." We are fortunate in having a cyclone cave or "hidey-hole" as it is locally known. Besides offering shelter during tornado warnings, it provides storage space for fruits and vegetables and canned goods. Dug into the side of a hill, it is dark, humid and with a temperature in hot weather that seldom rises above 70 degrees. During the winter months it is usually around 45 or 50 degrees; nearly ideal for growing mushrooms.

In one corner of this cement-lined cave, we have reserved a space about 3 feet square enclosed in 1- by 10-inch boards, treated to retard moisture absorption and rot. The average crop extends over a growing season of 4 to 5 months, and production varies from one pound to 3 or 4 pounds per square foot of bed. We do not try to grow mushrooms during the summer months when the temperature goes into the seventies.

Mushrooms must have a prepared medium containing organic matter and carbohydrates in which to grow. The original method was to compost horse manure and wheat straw. When a decline in the horse population made the essential ingredients difficult to obtain, it was found that a successful growing medium could be formulated from ground corncobs, cottonseed meal, hay, greensand and poultry manure. Variations of this formula have also met with success.

The horse manure is already well mixed with straw and hay when we get it in bushel baskets from a neighbor who keeps a saddle horse. To make mushroom compost, we mix this fresh manure with equal parts of well-rotted humus from our pit, add several handfuls of greensand, moisten it, and allow it to heat. Every 4 or 5 days the pile is turned and carefully checked by thermometer, and when the heating is reduced to a steady 70 degrees and the strong ammonia odor disappears, we fill the bed, packing it firmly but not too tightly, and keeping it well moistened but not soggy wet— there *is* a difference.

Care should be exercised in pulling or cutting the mature mushrooms to disturb the bed as little as possible. Mature mushrooms are from one to two inches in diameter and, despite all care, some of the smaller buttons will be disturbed, separated from their root filaments, and they will cease to grow. The bed should be examined daily and the dead buttons removed and used as they are discovered for, if left to decay, they may cause disease and kill out a portion of the bed. If you cut the mushrooms, the remaining stub or root should be removed within a day or two, for the same reason. Fill in fresh compost if there is a noticeable hole remaining.

Growth will be slower when the temperature goes down to 45 or 50 degrees, but the bed will last longer and produce a larger crop. Should the temperature go above 60, production will be speeded up but the life of the bed will be shorter.

—Victor A. Croley

Mushrooms may be cut at any stage close to the soil with a sharp knife. The gills begin to darken when fully opened, shedding spores that are not suitable for cooking.

... And Under the Kitchen Sink

Ever hear of growing mushrooms under the kitchen sink? I once met an elderly gentleman who did just that.

Living in a tiny New York City apartment, he nevertheless wanted to grow some organic food of his own. Herbs did fine on his window sills and he even grew radishes, lettuce and peppers in a couple of big window boxes. But his happiest experiment was with mushrooms. He found they thrived in trays under his kitchen sink, in the bottom of a large dish closet and in a spare bathroom cupboard!

The moral is this: mushrooms are a lot easier to grow than most people think. Provided their requirements of darkness, humidity and temperature are met, they take very little care and are great fun to grow—and even greater fun to eat.

Growing mushrooms is a kind of "reverse gardening." Being fungi instead of plants, mushrooms have no chlorophyll. Thus they are grown in darkness, having no need for sunlight to synthesize their foods. Instead, they rely on the richness of the compost on which their dusty black spores sprout and grow. A mushroom farmer wears a miner's lamp on his head, rather than a straw hat.

Homesteaders have found mushrooms an excellent spare-time crop to raise in barns, root houses, stables and chicken houses. According to the Ohio Experiment Station, growing mush-

Mix the dry ingredients thoroughly before saturating them with a cupfull of water.

rooms in winter in a house that is used to raise broilers in summer can be a high income producer. Poultry litter makes very good mushroom compost, giving higher yields than regular stable manure compost. This is probably due to its high nitrogen and vitamin B content.

You can, of course, buy prepared trays, complete with compost and planted spawn. Put in a cool, damp spot, they will bear a pretty good crop.

But if you have room in your cellar or other suitable building, you can start from scratch and grow a really big crop. The most important consideration is temperature.

The ideal air temperature is 60 degrees, but it can vary 5 degrees in either direction. Relative humidity must be high, as near 80 to 85 percent as possible. An inexpensive hygrometer will enable you to check this regularly.

Make tests with a thermometer and hygrometer in various spots to find those that maintain the right conditions day and night. Remember that there may be a difference of 10 degrees

or more at various levels in the same spot. There should be no drafts, but absolute darkness is not essential, as long as the light is very dim.

Usually, due to the difficulty of keeping the air cool in summer, mushroom growing is limited to the winter. But if you are lucky enough to find a spot that has the right temperature all year, you can grow crop after crop and get 6 or more pounds per square foot of bed.

You may find it most practical to make permanent box trays like greenhouse benches, setting them in tiers about a foot above one another. Or you can construct boxes without bottoms to stand on the floor, or close off a corner or end of the room with a single board on the floor. The beds should be about 6 inches deep.

Now for the compost. Fresh, strawy horse manure is the standby of commercial growers. It is piled up 4 feet high and turned, shaken and watered every 4 days. Dried brewers' grains, cottonseed meal or other nitrogenous materials, plus rock fertilizers, are

To keep soil from drying, completely cover it with plastic bag which fits inside band.

Home mushroom kit is ready to produce when the plastic lid is placed over the collar.

generally added to enrich it. In 2 to 3 weeks, when the temperature is down to about 75 degrees, the compost is put into the beds and planted.

Sometimes sawdust, shavings or ground cobs that have been used for bedding animals is composted instead of horse manure. Or you can make your compost with ground cobs or sawdust without manure:

Mix 100 pounds of the cobs or sawdust with an equal amount of straw, adding about 20 pounds each of tankage (or similar organic nitrogen fertilizer), leaf mold, greensand and granite dust and whole grains. Mix and tamp the pile well, water thoroughly and turn it in 5 days, then again in a week. Put it into the beds and plant when the temperature drops to 75.

Spawn generally comes in brick form. Break it into pieces a little smaller than a golf ball and plant it one to

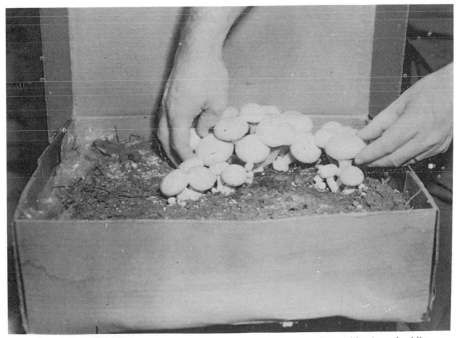

Despite inexperience, the beginner can achieve a mushroom harvest with or without growing kits.

two inches deep, 10 inches apart in both direction. Keep the beds moist but not wet; use a gentle spray of tepid, never cold water.

It takes about 3 weeks for the filaments or threads, called mycelium, to spread out completely over the bed. Then you "case" the beds, covering them with an inch of clean soil, free from any chemical residues. In another 3 weeks, tiny pinheads, the first "flush" or "break," will begin appearing on top of this.

If you maintain correct temperature and humidity from casing time on, you'll have no trouble with diseases or pests. A layer or two of moist burlap will help keep the proper humidity over the beds if the air tends to become dry.

In about 10 days, you can start harvesting your crop. But don't hurry the picking. Mushrooms are at their most flavorful when the bottom of the cap has broken away from the stem and spread out almost flat. Never leave them after this, as they stay good only a short time.

OKRA

Growing Okra In A Seedbed

Several autumns ago we started an okra seedbed by making a trench row about one foot wide by two deep into which we packed leaves, some manure, and a sprinkling of garden lime. The autumn and winter rains soaked into the ground and mellowed the mixture.

About one month before planting time, we take plenty of the leaf mixture out of the row and stir it into the soil at each side of the trench. We then convert the trench row into a compost tunnel, filling it with more leaves, gravel and wood chips, topped by a thin layer of soil. Later, if necessary, we add mulch to the strip to prevent loss of moisture and use the tunnel as an underground reservoir for maintaining moisture in the general area.

During the weeks to planting time, we stir the leaf-and-soil mixture several times to complete the composting. Before planting—usually the last week in April or the first in May—we again stir the leaves and soil thoroughly and, a few inches away from the seed row, rake aside any surplus undecayed leaves which might interfere with the young roots. Later, when the large, bushy roots develop, the composted leaves are mixed with the soil to maintain tilth. We have also found that okra grows better and produces more in soil that is slightly acid rather than alkaline.

To assure sufficient plant food for the young roots, we sprinkle a reasonable amount of rotted sheep manure compost (or any other good composted mixture) along the seed row and thoroughly mix with well-pulverized soil.

The pre-soaked seeds are sowed about 6 inches apart and covered with one inch of soil in rows 1½ feet from the trench tunnel at each side in order to give enough room for proper root development. Soon, in the early part of May, we have a "stand" of okra in our seedbed which we thin, especially where any two stalks are too near together.

Later, by transplanting, we thin the rows some more as we move plants to the areas between our early vegetables such as onions, which we plant in rows about 4 or 5 feet apart.

To get full benefit from our okra patch, we thin some of the seedbed, transplant if we have room, and leave some thickly growing plants—as close as 6, 8, 10 inches—to encourage earlier crops plus a continuous harvest. But, however you grow okra, you must maintain soil fertility and moisture so the plants can attain their full potential production. Okra should be picked at least every other day during the hot summer weather, with production slackening in the shorter autumn days.

Mulching—OK for Okra

In growing okra a good mulch is important if your soil is heavy and rain abundant. If you don't mulch, the soil must be loosened over the seeded rows with an iron rake as soon as it dries, while the space between must be cultivated to prevent a crust from forming.

Later, before the plants bloom, continue to work the aisles. Hill the growing plants and *mulch between the rows*

heavily with straw, old hay or well-rotted cow manure if it is available. *But mulch the rows lightly*—just enough to discourage weeds—because the plants need the warmth. Four inches of grass clippings are ideal. If rain fails, be sure to water well because okra is a great moisture consumer.

Okra, once it takes hold, grows rapidly and, before you realize it, blooms appear. Within 54 to 62 days after planting—about the time it takes snap beans to mature—you may be picking your first pods.

To keep the plants producing well into the fall, the pods must be picked every day or every other day. Picked when one to 4 inches long, they are at the acme of perfection: young, green and tender. If you pick them later, they are woody and no longer useful for cooking, canning or pickling.

Pods you cannot use or process immediately may be stored for a day or two in a cool room and sprinkled frequently with water to maintain a high humidity. Don't store them in closed boxes or crates because they will heat up and burn beyond use. When you store them in a closed space, be sure there is good air circulation to keep temperatures down.

If some pods get ahead of you, shell the matured but unripe seeds and cook them just as you would green peas. If you want to experiment, let the seeds dry, grind them up and use them as a coffee substitute.

—Betty Brinhart

Pruning the Pods

Pruning is the secret of good pod production and is started when the first blossoms appear. Begin by snipping off about one of each 3 leaves, and continue the process throughout the season. This causes growth to go into fruiting and makes the plant branch down low. To continue this process of outward growth, prune more heavily on the inside of branches, and prune more heavily during prolonged wet spells.

This method will produce plants about 5 feet high with about 5 main branches, bearing early and continuously until frost. When the fruit size gets too small in late fall, prune every second "bush" severely and thin the others, leaving enough buds to supply the table. Give a supplemental feeding of well-balanced organic fertilizer if leaves start to turn yellow.

To make certain the gumbo has a continuous supply of food (it's a very heavy feeder), I mulch with straw or hay and add compost or manure as available by spreading it on top of the hay as a top-dressing.

You can grow an early-season crop if you should prefer to follow okra with another crop. Grown my way it is the most reliable vegetable in the garden since it isn't bothered by diseases and bugs, and the withering heat speeds up pod production. Two 30-foot rows give us plenty for eating fresh, giving away and freezing.

Known generally as a thickener of soups, okra is also tasty and useful to those who prepare it as a cooked and salad vegetable. It is high on the list of vegetables rich in minerals and vitamin A, with a half cup containing about 500 units. Try growing and enjoying some yourself.

—Jim Evans

ONION

A Tearless Way

Over the years, my family and I have experimented with growing onions in ways that depart from standard methods. Being great lovers of onions we grow and use quantities of them. Our method for bringing them to early maturity will undoubtedly prove of interest to many gardeners.

During the first week of April we plant onion sets in a raised bed of deeply-dug, compost-enriched soil.

Manure does well in place of compost. (Over a number of seasons my sons and I have created the raised bed. It consists of numerous barrowloads of good garden soil dumped in a suitable sunny location. The bed rises a foot or more above garden level. It drains beautifully, warms well ahead of surrounding soils, and allows far earlier working and planting than the rest of the garden area.)

98

With the raised bed worked to a fine tilth, we space the onion rows about a foot apart. That's because the boys and I have big feet and it would be impossible for us to work in narrower rows without crushing the plants. People with daintier feet may well use slimmer rows, say 9 or 10 inches wide.

It is a common practice to make furrows for the onion sets. If the furrow is no more than a mark to indicate the row, fine. But if the furrow is deep, then the sets will be deeply buried, a bad situation since onions are surface growers. The sets should preferably be thrust into the soil so that the tops peek out. They need plenty of light and air for rapid growth and good development. If onion plants are used, plant just deep enough to hold them erect. Planted too deep, onion seedlings will mature into thick-necked, small bulbs which will prove to be poor keepers. Sure, the earthworms and rains may push such shallowly planted sets or plants out of the rows here and there. But it is the work of only a few minutes to thrust them gently back in place.

Because the raised bed warms so quickly and drains so well after torrential and frequently cold early-spring rains, the onions are off to a head start of 5 or 6 weeks over those planted later in the main garden. Thus we're chomping green onions while most gardeners are just getting ready to plant theirs. The ones grown in this fashion are of good size and quality, too.

We plant sets or seedling onions about 4 inches apart in the rows. But we have found that thicker plantings made in composted or heavily manured soil will result in onions of good size. When purchasing onion sets, only those about the diameter of a dime should be chosen. Contrary to some beliefs, larger sets will not develop into larger onions. Many go to seed instead.

Soil should never be drawn over growing bulbs to cover them. Onions feed from the shallow roots reaching through the surface of the soil. Needing much air and light, onions that are covered do not ripen as quickly or store as well as those exposed to these elements. In areas where there is plenty of rain it is not necessary to mulch onions. Hoeing carefully will eleminate weeds, but care must be taken not to disturb the soil in which the roots are anchored.

If weeds appear among onions in the rows, wait until after a rain before eliminating them. They should be pulled out gently by hand, disturbing the plants and bulbs as little as possible. Where long dry spells prevail, mulch the onions to conserve needed moisture. If the soil should dry for some time, the onion will form two small bulbs or "splits" rather than a single large one, as a result of the lack of adequate moisture.

The mulch we favor most is old hay. It is easily available and cheap. A very thin scattering is made as soon as the green spears appear. Additional thin layers are added as the onion stalks grow in length. This permits the onions to grow well through an easily-penetrated mulch, and allows both light and air to reach the bulbs. Too heavy an initial application of mulch will result in loss and anemic growth. If it appears that the mulch has become too compacted, fluff it with a hayfork or a steel-tined weeding implement.

When the tops have broken and fallen indicating maturity, a rake or broom may be used to flatten all stalks still standing. When the stalks show a loss of color, the onions should be pulled and spread in the sun to cure before storing.

Raising onions in a constantly replenished mulch will make them grow very slowly. We always put in several rows like this for the succulent greens they produce over a long period. It is true that the matured bulbs are mostly of small size and often thick-necked. They do not store well at all and hence must be used as soon as possible.

A small planting of onion sets mulched with wood chips proved interesting. The stalks grew thickly and heavily and slowly. Again we enjoyed a long period of succulent onion greens. But the bulbs scarcely widened out at all from the thick neck. Nor were the onions grown in wood-chip mulch for storing. They had to be consumed in short order.

—John Krill

Prize Onions Mean "Something's Right"

"When you raise onions that weigh from 1½ to 4½ pounds each—and the whole crop turns out that way—" says Mary Louise Leidle of Malibu, California, "—you must be doing it by creating conditions that agree with onions."

Miss Leidle's giant Sweet Spanish Hybrid onion crop—several 50-foot rows of beautiful onions—was obviously prize-winning from the start. The onions grew fast, were well fed and irrigated properly, and completely justified faith in organic gardening.

"From the start we have used only natural manures and wastes," says Mary Louise. "We brought in truckloads of bean straw, spoiled hay, horse and cow and chicken manures and

piled this thickly over my 50- by 100-foot garden space. I soaked it deeply with the hose. When the soil was workable, I turned these materials into the top 8 inches with my rotary tiller. Then another soaking and a working with the tiller.

"I do this yearly. After the last tilling, I let the soil rest for a month to digest the fertilizer. By that time the weeds have started. I turn them under and the soil is ready for planting."

Although onions can be planted about any time in California, Miss Leidle says her methods can be applied anywhere. About two months before she wants onion plants to set out in the main garden growing rows, she prepares small seed flats, fills them with a mixture of 60 percent clean sand plus sifted leaf mold, and plants her onion seeds in this firmed-down mixture.

"I always grow my onions from seeds. They seem to do better than sets. You don't have a choice of many fine varieties when you buy just red or white sets. This crop I raised was Sweet Spanish Hybrid. They grew big, were mild and sweet, sliced divinely, were excellent in relishes, etc., and, according to the seed growers, are good keepers."

Miss Leidle usually plants her onion seeds about October 1. The seeds may be either broadcast in the flat or planted in rows. The small plants will be about 4 or 5 inches tall by mid-December, and ready to set out. The growing rows are 18 inches apart, and the sets are planted two together and 12 inches apart. Later the weaker of each pair is pulled out, thus leaving the huskiest onion to sprint the rest of the way down the growing lane to harvest time.

As to proper curing: Miss Leidle lets her onions grow to complete maturity in the ground, until the tops wither over, fall and are dry. Then follow several days' curing in the hot sun atop the growing ground in the rows.

"A lot of onions will spoil on you," warns Mary Louise Leidle, "unless they are properly cured. Never harvest immature onions. When stored they must be kept in well-ventilated containers—preferably loosely woven bags. Proper storage temperature is above freezing and not over 50 degrees. A cellar is good—dry and dark. Hang sacks from the ceiling. Also pick a good-keeping variety: the Bermuda type onions are poor keepers; the Sweet Spanish are rated good keepers."

—Gordon L'Allemand

Spanish Onions—The Secret of Growing Them Big

What we do to get really large onions is to dig a 4 inch deep trench for the plants then fill ½ of its depth with sifted compost. We cover this with one inch of garden soil, then set the plants.

After working your bed, wait for a good rain to settle the soil. Set the plants about 6 inches apart in rows at least 12 inches apart. Remember these grow twice as large as ordinary onions, so need more room for weeding and cultivating.

Do not plant seedlings too deeply. Cover lower ½ inch with soil, then firm earth very well around each. If rain isn't in sight, water entire bed after planting.

Spanish onions need only shallow cultivation. Loosen soil down to 6 inches with a grub hoe after each rain so roots will have good air circulation.

After 4 weeks these onions may be pulled and used as bunch onions. Mulch rows now with 3 to 4 inches of any available material. Water well if rain is sparse. Since they develop into big fellows, Spanish onions need plenty of moisture. If soil remains dry around roots for a prolonged period, onions develop the "splits," each becoming two smaller bulbs instead of one large one. Remember, too, that sunlight influences size. Spanish onions need at least 14 hours a day; plant them in the sunniest part of your garden.

Spanish onions mature in 110 to 114 days, ripening around the last of August. Pull soil away from bulbs so upper half of each is exposed to sun and air, which aids in maturity and good keeping qualities. When most tops have fallen over, push the rest down with a rake. A week later, pull up the onions and allow them to remain on the surface for 3 days to cure in the sun. Outer skins will harden, thus preventing rot in storage. When dry, snip off the tops one inch from bulbs, then place in an airy outdoor building to cure until cold weather sets in. Don't braid stems as for other onions; these are too heavy and will soon break off.

—Betty Brinhart

Onions That Don't Know When to Quit

Last summer I raised onions that were the biggest I had ever seen. I set them out in March among my roses and strawberries. It was very cool all through the spring, and we had quite a bit of rain for this part of southern California. When they finally got started, they didn't know when to quit. Everybody who saw them marveled at the size of them. So one day

—the 14th of June—I decided I would pull 10 of the largest of them and take them to a market close by and weigh them; I knew their scales were correct. To my surprise they weighed 23½lbs. *Unbelievable,* but it is a fact.

I have always been organic-inclined and now I know it pays off. I have a 45-foot row of roses 33 inches wide. And in that small space I grow roses and strawberries, onions and rhubarb enough for my wife and me—and have some onions and rhubarb to give away.

I have a compost pile that everything that will decay goes into. I even bring sawdust from the railroad to add to it. I have my soil built up to where I have the most beautiful dichondra lawn on the street. People want to know what I feed it. I just have to tell them it is the soil and plenty of water. —Charles A. Harkey

Onions That Grow on "Trees"

We've found a perennial vegetable you ought to know about. It's the Egyptian onion—often called a tree or top onion—which produces small, tangy bulbs at the tips of its flowering stalks. Not only are they delicious eaten fresh or in salads, but the plants require practically no care at all.

Perennial onions may be sown from seed in the spring, or planted from sets during spring or fall. We've found it relatively easy and less expensive to plant Egyptian onions from bulblets or sets. The small bulbs can be planted almost any time during the growing season, but don't expect much the first year, as it takes a full season for bearing-size plants to develop. The sets were initially planted about 4 inches apart in compost-rich, well-drained soil with about a foot between rows. It wasn't long, however, before we transplanted the seedlings around fruit trees— where they bear just as well, take up even less room, and aid in the war against insects.

Cultivate top onions much the same as ordinary types. Work manure or rich compost in around the plants each spring, and keep the soil moist with a mulch, as no onion will thrive in dry sites. We never cultivate more than a few inches beneath the surface of the plants, since the onion is a shallow-rooted crop despite its being a bulb. Additional top-dressings of organic fertilizer during the growing season help production.

Our Egyptian onions reached a height of over two feet the second season, 10 mature plants yielding about 30 portions over a 3-week harvest period. An easy method of propagation is to plant the small bulbs that form at the top as soon as they break from the swollen stalks and begin to mature in July. Divide plants every three years or so, breaking up the large clumps of underground bulbs into smaller clusters and resetting them. That's a task we look forward to next spring—and to our crop of onions that grow on "trees"!

Editor's Note: Sources for Egyptian or multiplier onions include Greene Herb Gardens, Greene, Rhode Island 02827; and Nichols Nursery, 1190 N. Pacific Hwy., Albany, Oregon 97321.

—Robert Hendrickson

The Space-Saving Onion Patch

Because onion cookery is a favorite at our house, I wanted to grow more onions than I could find room for in the vegetable garden. I decided to give close planting a try. I was able to spare a section about 40 inches wide between the future cucumber bed and a row of peas—and decided *this* would be our onion patch!

The ground was spread with aged horse and chicken manure, then sprinkled with wood ashes and bone meal. This was worked in and the surface raked smooth. Marking off 5 rows approximately 10 inches apart with the hoe blade, I set in two pounds of yellow Ebenezer onion sets, with the help of the children—who bend over so effortlessly. Keep the sets two or three inches apart in the row, and discard any smaller than dime size--they lack vigor and will not produce a good bulb. I covered the sets with about one-fourth inch of soil, making sure each one was in an upright position, then firming the soil around them well. This takes a little time, but it's well worth the effort for a good stand of onions.

One small reminder from experience: Check your onion plantings every other day for a while to reset those bulbs overturned by night crawlers—plus the few that are pulled up by birds. In my garden, the night crawlers sometimes come to the surface too close to an onion set, and when they reverse direction to submerge again, they nearly turn the onion set upside down.

Outside of an occasional weeding, this closely-planted patch received little attention. I mulched with spoiled hay around the patch in June, and during a hot dry spell in July, put a light covering of mulch between the rows.

By the end of August, the onions had matured, their tops nearly all having fallen to the ground. The remainder were toppled by the children. and a few days later, I pulled the bulbs, piling them in a row to cure

After their tops are cut, onions are gathered and stored in a cool place to be cured.

in the wind and sun. It is recommended that several days after this, the tops be cut off about an inch from the bulb and the onions spread loosely in a shed to complete the curing process until cold weather—then be stored in airy containers in a cool, dry place. Last year, however, I left the onions spread about in the garden until the tops were completely dried, then stored them in a basket on a high, dry shelf in our basement—where they kept beautifully all winter.

Since the number of daylight hours the onion set receives each day determines when it becomes an onion, I plan to get mine into the garden as soon as possible in early spring, which here in Maine is some time in late April or early May. The Ebenezer and Yellow Danvers set which I use require 13 hours of daylight each and every day to bulb. Others, such as Yellow Bermuda, require 12 hours daily, while the Red Wethersfield needs 14—but the red variety is the least likely to be damaged by the onion maggot. I've found that late frosts, which we get right until the middle of June, or even a light snowstorm, doesn't keep the sets from growing and doing well. So I put them in at the same time I plant carrots and peas.

If your plantings are bothered by the onion root maggot, it's one sure sign you need to add organic matter to your soil. Radishes interplanted with the onions will also help trap these pests, and they can be pulled and destroyed when infested. If thrips are a problem in your area, plant the Spanish onions that show considerable resistance to this garden pest, which causes abnormally thick necks on the stem of the plant, preventing normal development of the bulb.

Although many people with whom I've talked seem to think that onions are difficult and not worth bothering with in the home garden, it's been my experience that they are relatively free from plant disease and are easy to grow. If the soil has had plenty of

A healthy stand of onions is flanked by late peas on wire supports while on the right the cucumber bed sports mounds of mixed nasturtium-and-cucumber plantings.

organic matter incorporated into it, and the area limed when needed, I see no reason why this popular vegetable can't be grown successfully here in Maine as well as in other parts of the country.

From that spring's closely-planted onion patch we harvested nearly two bushels, plus all the young onions we enjoyed during the summer. Although they were not the spectacularly large bulbs raised by some organic gardeners, they were nearly all a fine size, and the best harvest I've ever had. So I shall continue this space-saving onion patch from now on—and certainly recommend it to anyone whose garden plot suffers from "growing pains." —Wanda Ferguson

Stalk Tipping Signals Harvest

Harvesting begins as soon as the onion stalks tip over. These are pulled before a rain so as not to stimulate regrowth, and to maintain their keeping qualities. Of course, too wet a season can affect how well they'll keep because the onions won't cure out as well.

When the onions are pulled, I let them dry outdoors first, then take them inside until the tops have completely dried. Once these are cut off an inch above the bulb, the onions are then ready to be stored in a cool, dry place.

I pull onions only when the tops have tipped over (caused by ripening or drying in the neck region). The rest are left to grow until their tops also fall over. I harvest from the first tipping until freeze-up time when I take in any still in the patch. Some tops stay thick and green and never topple. These are kept out to be used up first because they may not keep in storage.

—Clara Fessenden

PARSLEY

Grow It Indoors

Parsley can easily be grown in pots on a sunny sill for winter use. Fill 6-inch clay pots with a mixture of half compost and half garden loam in August, and sow 5 or 6 parsley seeds. Keep in shade and do not allow to dry out. They can be sunk in the earth in the moistest part of the garden. Before the pots are brought indoors, on a warm October day, the seedlings can be thinned to two or three in each pot. Potted plants may also be started from seed indoors during fall or winter months.

Indoors, parsley does best in a cool room on a windowsill where it receives plenty of sun. It benefits from a weekly sprinkling of room-temperature water, but can be put outside during warm winter rains. Liquid fish fertilizer can be given every two weeks, or water in which raw eggshells have soaked for a day or so.

You Can Grow Permanent Parsley

Select for your parsley bed a portion of the garden which can be given over to it permanently. Then divide this area into two sections. Plant one with an annual crop the first year so that this part of the plot will be empty again by fall. Plant the other with parsley which, incidentally, germinates more rapidly if you soak the seed overnight, and water the planting through a single layer of burlap. At the end of the season leave the parsley plants in place. They will come up early in the spring and provide you with a lush harvest until summer. Then they will flower, and no matter how often you cut back the flower stalks they'll get ahead of you and insist on going to seed. Don't argue with them. They are making your next year's crop. This year, the second planting year, you will have sown new parsley in the second section of the bed, the one that contained an annual crop the first year. Or perhaps you will want to select an entirely different portion of the garden for this second planting. In either case, by the time last year's parsley plants have gone to seed, the second crop will be harvestable.

The second fall just pull up the old plants from the first year, rake over that section of the bed and leave it alone. Next spring it will be full of young parsley seedlings, volunteers this time, and the cycle begins once more. The second plot will take care of your needs for a while, then go to seed—for the following year. The volunteer seedlings from the first year's planting will make the third year's crop. (Although they will come up in an approximation of a row, you will probably have too many of them, but there's not much work to thinning, after all.) And so, barring accidents, it will continue, year after year. From now on you can forget the seed-soaking and the burlap. Your parsley will take care of itself.

—Elizabeth Rigby

PEA

All Kinds Are Great

The secret with peas is to plant early. Peas do best in cool weather. In fact, in the South, they're usually a fall crop. There is good reason for the tradition of planting the first peas on St. Patrick's Day. It's exactly the right date here (near Boston) and in most northern areas. Plantings a few weeks later may germinate better and come along faster, but they never quite catch up.

If the ground can be worked— which with me, means merely drawing aside the winter mulch and making the furrow—I sow the seed in mid-March. Frost or snow later on does not harm peas. I make succession plantings every 10 days or so up to May first.

I harvest the first picking on Memorial Day or a little later. Peas that mature in hot weather deteriorate in quality, so I don't plan on any after the fourth of July.

Besides working in organic materials in fall, I spread wood ashes from my Franklin stove on the area where peas are to go. Wood ashes contain some potash, which peas need.

Before planting, I open up the furrows to warm up a day or so in the sun. Remember, the ground has been covered with a thick mulch and the weather may be almost freezing. I make a deep furrow rather than a "trench," sow the seed about an inch deep (later plantings deeper), so the roots will keep cool. Directions on seed packets usually say to sow a dozen seeds per foot. That means later thinning if the seeds sprout well, which mine do. I sow seed two inches apart and rarely need to thin out.

If you have finished compost, spread a layer in the furrow before planting, then another layer over the peas as a cover instead of ordinary soil, firming this layer down well. Or you can use dried manure. I bought mine—a dried cow manure compost— at a supermarket; it's completely organic. This is all the fertilizer peas will need —although I should add that my garden gets rotted cow manure from a local dairy spread on in January. But I don't waste much on the area planned for peas. Other crops, like tomatoes, squash and melons, are far more demanding.

After planting, I draw back the winter mulch just to the furrow. It could cover the latter loosely. Peas are a coarse plant and will come up through it.

Most dwarf peas, as well as tall varieties, need support. Usually I set up chicken wire for the tall types, and branches of deciduous trees for the dwarfs. To save time, cut the brush before spring. Set supports at planting time. One line of brush will suffice for a double row of dwarf peas, spaced about 8 inches between rows. Between each set of double rows allow about two feet. Space tall peas correspondingly farther apart; seed packet directions usually will provide distances. As seedlings grow, fill in furrows about the base of plants.

Do not begrudge the space peas take up. They're out of the way relatively early, and the area can be used for many things—carrots, beets, beans. Seedlings of lettuce, cabbage, cauliflower, broccoli or fennel for fall use can also be set there.

Moreover, peas leave the soil in good shape, unlike the greedy crops. All legumes have a beneficial effect. And thanks to the winter mulch, which has just about disappeared by the time the peas are pulled up, the vacated space is practically weedless. It's mellow, too. You won't have to dig it up for the second crops. If you can, renew the mulch; use grass clippings, hay, straw, weeds or the nitrogen-containing pea vines themselves, which also make a valuable addition to the compost pile.

Grown under proper conditions, peas are little troubled by disease or pests—at least mine aren't. Once in a while, a cutworm severs a seedling, or rabbits nibble the sprouts. (They prefer lettuce or cabbage.) The mulch of organic material in my garden probably prevents root rot; peas on poorly-drained land are susceptible. Aphids may attack peas; here again poor culture would be responsible.

—Ruth Tirrell

Electroculture for Extra Harvests

Edward P. Morris has reported how he used "poultry mesh or welder wire fencing two feet wide by 40 feet long for training dwarf and early peas" in his Ames, Iowa, garden. After erecting his fence, he lifts the young vines and trains them to the wire fencing using laths until the "tendrils take hold." The results amply rewarded

him for his careful planning and work. "It was a wonder to see! I harvested 3 separate pickings from the same vines."

Looking along his 40-foot wire fence he saw "the tops of the vines were one continuous bloom for the full length of the row." Electroculture, many of our readers would say, had something to do with such excellent results.

Double-Row Pea Culture

I grow my green peas in double rows, back-to-back, staked and wire-supported two to three feet off the ground.

In addition to producing healthier plants, this method makes feeding, cultivation and mulching easier. Picking, once a tedious family chore, is now a pleasant operation because you don't have to bend over so much and the pea-filled pods hang where you can see them without peering and hunting.

I plant drought- and wilt-resistant varieties, using both the dwarf and taller late producers which permit successive crops plus interplanting with corn and tomatoes. Because peas prefer cool weather, they should be started as soon as the ground is workable. We keep the soil well-prepared in advance, rotary-tilling green manures, leaves, rock minerals and old mulches ahead of the season. We also sprinkle ground limestone every 4 to 5 years as a blender of soil nutrients.

Peas are a short-time crop, and since I want to keep my garden working all the time, I plant them in twin rows 8 inches apart, leaving 3½ to 4 feet between the double rows so corn and tomatoes will be able to take over when they are through. When the peas reach the tendril stage—8 to 10 inches tall—it is time to start staking

The tall end stake supports and holds the stock wire between the twin rows of peas.

Feeding between the rows of young plants includes bone meal and granite or compost.

Wire is unrolled before being erected between the rows held apart by wood laths.

Insert a whip to hold vines against wire, and reinforce structure where necessary.

Remove laths from the base of the vines as they grow and tendrils take firm hold.

Bone meal plus minerals are hand-worked into the soil between twin planting rows.

Fertilizers are tilled into the soil outside the rows while laths are yet in place.

Hay mulch the entire area 3 to 4 inches so the soil is covered up to the plants.

them out. Here are some of the materials I use, but it is a good idea to work with whatever comes free and handy—the important thing is to get your peas up off the ground.

1—For the dwarf and early peas, I use poultry mesh or welder wire fencing two feet wide by 40 feet long. I also accumulate odd pieces of wire mesh of varying lengths and widths for the taller hybrid peas.

2—I always keep a supply of stakes on hand, about one inch thick and 3 or more feet high. They can be old broom handles, tree, bush and brush prunings which I sort according to size and then tie in bunches. I also keep a collection of dry laths for marking and general use in the garden, including pencil-thick whip cuttings taken at pruning time.

3—I collect and save old or spoiled hay for mulching. When I use hay or straw for mulching between rows, I add cottonseed or soybean meal.

Using the laths, I separate the pea rows, bending them out and away from each other until they rest on the ground which opens up the 8-inch-wide center strip for feeding and cultivation. After cutting the weeds with a hoe, I sprinkle the area with bone meal, ground granite or compost, working the additives into the soil and taking care not to disturb the roots.

Between the twin rows, I erect the wire fence, starting at either end and using the wooden stakes to anchor it in place. I work or weave the stakes through the wire strands, thrusting them one foot into the ground, and spacing them at 3-foot intervals. If necessary I reinforce the structure by weaving the wire and wood together, using whip cuttings at the critical places.

Taking care not to injure the delicate young plants, I slip the laths behind them and lift the vines up off the ground and against the wire. I keep the laths in place by thrusting several sticks behind them into the ground. Later, when the tendrils take hold, I carefully remove the laths and store them for subsequent use.

After working along the second twin row in the same way and securing the vines, I side dress the rows, working bone meal and granite grit into the soil with a rotary tiller. Finally, I hay mulch the entire area to a depth of 3 to 4 inches, completely covering the soil right up to the plants.

The same method can be used for the late, tall varieties, with 40-inch stock wire and longer, heavier stakes set 4 to 6 feet apart, depending on the length and height of the row. The most important part is the careful handling of the tender young pea vines so they are not injured or torn. Again, laths may be removed from behind the plants when the tendrils take hold of the fence. I dry them out carefully before storing them, so they can be used again—which is the way to take care of all the wooden implements used anywhere in the garden.

The patch is now left alone to bear and produce until picking time. It is good to check the rows, and pick an occasional weed thrusting through the mulch and to secure sagging parts of the vine with the whip cuttings. When the vines are through bearing, they ᴠre pulled up and used on the spot as mulch for subsequent plantings.

—Edward P. Morris

Peas Can Be Pretty— And Practical!

Mud sliding down a newly excavated slope is a discouraging sight, so early one spring I planted dwarf Chinese peas to hold my garden together until I could plant permanent ground cover.

I planted peas because they grow quickly in cool weather, their roots stretch tenaciously, and they also improve soil. Being legumes, they are workers with nodules at the ends of their roots which add nitrogen to the soil. This helps fertilize shrubs and trees nearby and prepares the soil for a lush future for some leafy plant. Plants growing adjacent to peas often do better, I have found, than they do by themselves. This, of course, is partly true because these plants also profit from water and fertilizer given to peas.

But, in addition to these excellent if practical reasons for planting peas in a "tough" place, I suddenly became aware of an unexpected bonus —people began complimenting me on the appearance of my former mud patch. They mistook the two-tone purple blossoms for sweet peas, and I realized that edible peas may be used satisfactorily as ornamentals when a garden is too small for a vegetable plot. Now I could enjoy a highly decorative floral vine that would allow me to pick pods instead of blossoms.

While edible peas don't offer the vast color range of the sweet pea, the climbing varieties will cover a trellis with a pleasing spring-green coat, while the dwarfs make good edging plants or temporary ground cover. And it must be conceded that the edibles have not been especially bred for ruffles, for fragrance, or for large, long-stemmed multi-blooms. But their blossoms are either purple or white, and you can plan for them in advance because the whites come from cream or green seeds, and the purples from gray or brown seeds.

The vegetable seeds are larger and have softer shells than the floral. Those with wrinkled seeds are sweeter but less frost-hardy, and are called late peas, while smooth seeds are better for winter or for the far North. The vegetable ones germinate a little faster and usually mature sooner than the floral.

The floral species can be differentiated by their seed color too, although it takes skill to be precise. Dark seeds produce red flowers; light produce white; and mottled, purple. But the in-between colors are hard to predict. Dark seeds germinate slower and need more water. Soaking them first in cold, then in hot water, helps crack the shell. Light ones sometimes rot in too-wet or too-warm a growing medium.

What I accomplished in holding the soil that spring could have been achieved as well with dwarf sweet peas such as Bijou or Little Sweetheart or with the low-growing, shelling kind, Blue Bantam or Little Marvel. I choose Chinese Dwarf Gray sugar peas because they are seldom available locally, and when I do find them, they are expensive or old. Chinese peas are eaten pods and all, especially in oriental dishes. If they stay on the vine too long, they can be shelled. Growing them not only gave me a convenient source—they freeze well—but I could have them fresh-picked just a few moments before cooking.

Getting peas fresh, at the "fleeting moment" of their perfection, as someone once said, is the best reason for growing them. As with corn, the sugar in peas begins to turn to starch shortly after picking. Peas also become starchy if they are left too long on the vine. Edible pods should be picked neither titmouse small nor big and bumpy. The shell varieties should be bright green and sugary instead of yellowed and starchy. Bite into several raw ones at various stages and taste the difference. When you grow your own you can be selective.

Freshness is reason for growing the floral species, too. Fresh-cut sweet peas last longer and are much more fragrant than those standing in a florist shop. The selection of colors offered by seedsmen is enormous. You can choose winter, spring or heat-resistant summer sweet peas.

Neither the edible nor the floral are much work, provided they are grown at the right season. They are winter, early-spring or summer crops depending on where you live, preferring the season just after freezing and just before the predominantly 80-degree, dry sunny days. Light frost doesn't phase them, though they won't stand ice-forming freezes.

—Marguerite E. Buttner

A "Sunpole" System for Pea Stakes

I enjoy growing sunflowers, but one fall, after removing the seed heads, I left the stalks standing. I planned to cut them down later. But, well, time passed and fall set in. One day I again looked at the stalks and reminded myself that I had better cut them down. Suddenly, and quite by chance, it occurred to me that there, as nice as you please, were my next spring pea stakes! Why not? So I tied them into a grid using sunflower stalks. I had two parallel rows, and there were two plants per hill.

Tied up into "tepees," sunflower stalks make effective supports for spring peas.

The grid withstands the rigors of winter and next spring the peas entwined the sunpoles quite handsomely. There are two advantages—no poles to store and no poles to replace in the spring.

—Joseph L. Horvath

Raising Peas on a Spruce Trellis

We train our pea vines to grow on dried spruce boughs because they are handy, free for the taking, and do a first-rate job. For years now, we've been using them to raise our tall Telephone varieties up to 8 or 9 feet, using stepladders when we pick them.

The peas are planted in double rows, 10 inches apart, when the seedlings are only 4 to 5 inches high—before they are tall enough to be torn or injured—*we insert the dried spruce boughs between the double rows so the vines have something to climb on.*

Strong winds here on the Maine

A good stand of high, sturdy brush can be worked easily into a serviceable palisade.

coast play havoc with poorly set brush, so a weaving of big spruce boughs goes in down the line, every third or fourth hole. The first inserts are big, sturdy affairs—9 to 10 feet long and up to two inches across the butt. We push or tap them 10 to 12 inches into the earth because they must provide the sound framework around which the whole row of brush supports is built.

We take a crowbar and make 10-inch-deep holes, 16 inches apart, down the entire row and between the double files of plants. Next we thrust a spruce limb as far as we can into the end hole with its straight side toward the path. Then, working from the other side of the row, we stick the next bough 3 or 4 holes away so it catches and braces the edges of the first bough. All curved boughs should face inward, toward the center of the double row.

The slighter spruce limbs—still 9 feet high—go into the holes between the heavier poles, all curved in to the center. These smaller units, interwoven both into the larger ones and each other, form a tight latticework of branches which encourages our pea vines to climb. We fill the slightest opening at its base with the smaller branches to make a brush curtain of rough, twiggy surfaces to which the pea tendrils cling readily. When the vines grow high and heavy, we anchor the entire structure with binder twine, tied horizontally.

We follow the same procedure for our sweet peas which we grow in two 40-foot double rows every season, using the same length boughs because our sweet peas grow just as high as our edible peas.

Thanks to carefully calculated spacing, the lofty framework does not shade the rest of the garden or even the adjoining pea rows. We usually grow 5 double rows of peas, each 40 feet long, and 5 to 6 feet apart, which allows ample space both for sunshine and moving about during picking. The peas appreciate the shade and cool, and are safely out of the way by the time the rest of the garden is fully mature.

Admittedly, the towering curtain of brush *does* seem absurd when the tiny pea seedlings are a mere 4 inches high at its base. But the peas are not discouraged. The latticework seems to act as an incentive because the active little fellows proceed to scramble up the palisade until they reach its very top where they may be seen waving in the breeze, looking for more spruce boughs to scale.

—Helen Nearing

Sunshading for Two Crops

Last summer I did a little "experimenting" which turned out well. I have just a small garden—about 20 by 12 feet—in which I devote 6 rows to peas, setting them in 3 pairs with walking space between. The two rows of each pair are kept 8 inches apart.

Before planting seeds last season, I stretched two-inch chicken wire, supported by 4-foot steel fence posts, down the middle of each two rows. Now, my home is in southeastern Wisconsin on Lake Michigan's shore, where the breezes come from all directions; so, as soon as the young peas grew high enough, the winds from each side blew them against the chicken wire and their tendrils caught on tightly. Of course, I helped a few that didn't cooperate. The vines grew

Down in Virginia, tender crops are protected from the midsummer sun by a lattice of laths.

strong and the pods hung free. Stormy winds didn't flatten the plants as had happened in other years.

When the last peas are picked and the vines pulled up, I've always bewailed the empty space in my garden. But not last summer. I tried another new idea. Since most summer nights are cool here along the lake, I reasoned that if I could restrain the hot August sun in the daytime, I might be able to coax up another crop of peas. So, in went the left over seeds. In just a couple of days those peas arched their necks out of the ground, then straightened up and popped their heads out. Immediately I gently tucked dry grass clippings around each plant and mulched between the rows deeply. They got one last thorough watering before I shaded them.

Three 12-foot two-by-two's were attached to the fence posts at the top of the chicken wire. They became the support for two lengths of snow fence laid flat over the top and fastened down with wire. The slats of the snow fence produced stripes of shadow which moved across the garden as the sun moved across the sky, breaking up the heat. It looked promising.

Then came a really hot streak just a few days before I was due to leave on vacation. Even the "filtered" sunshine was too much for my seedlings. So I brought out a couple of old bedsheets, stretched them over the top, and fastened them securely to the wire of the snow fence with nails. That did it! When I returned home after a two-week hot spell, I found the peas grown tall and fresh. The earth beneath the mulch was still moist.

From now on I plan to have two crops of peas every summer!

—Marjorie Dee

Fencing is easily handled, and can be used in winter to protect work areas or delicate plants.

PEA

Peas in the Pod

Ever start out with a dishpan full of freshly picked peas and wind up with a dishpan full of empty pods that looked good enough to eat? It seemed almost a shame not to, they were so crisp and fresh and juicy, especially if they came from your own garden. One look at the few handfuls of peas in the pot, all that remained of what looked like a big harvest, and you couldn't help wishing all the more that you could cook the pods, too. But even though Nature does gift-wrap the pea you just can't eat the wrappings.

Not, that is, unless it happens to be that most delectable legume, the sugar pod pea, the *mange-tous*, or "eat-all" as the French call it. It's not a bad name, for that is exactly what you do, eat all, pods included. It's an old favorite in many country gardens, but most people who must depend on city markets for their fresh vegetables have never even heard of edible-podded peas.

Sugar pod peas produce most abundantly and, not only is there no "waste" to your crop, in the form of the usually discarded pods (although you do add them to your compost pile, of course) but each plant bears so heavily that a pan full can be picked in no time.

It's a good idea to select one of the cooler spots in your garden for the peas. Avoid a southern exposure. In the case of spring plantings, if warm weather arrives prematurely, the vines will bust out in bloom and you'll have to hurry and pick a nice early crop, but it will be a smaller one than you'd have at the end of a *longer* cold spring. Now, by careful timing the fall planting can escape this hazard. Just be sure, however, to consult the frost schedule for your region before planting and allow about 65 days for your plants to mature. The cool of the fall can be favorable to pea culture provided care has been taken to plant far enough ahead of the frost date line.

Once the flowers begin to fade you must keep a close watch to catch your peas at the peak of perfection.

The picking must be done with great sensitivity, a fine feeling for that just right moment. A day later and the flavor is overblown. In the case of the sugar pea there is a much wider time margin here. The fact that sugar pods can be eaten long before the peas inside have fully developed makes them available considerably earlier and for a longer period of time than ordinary peas.

Incidentally, the tenderness of the pods of the sugar pea is due to the absence of that parchment lining present in the pods of all other kinds of peas.

Sugar pod peas are planted just like other kinds of peas. Being so prolific they are a good space-saver, something to consider with all the slow growers like corn, tomatoes and melons taking up so much room in the garden.

—Clee Williams

PEANUT

A Sure Crop

"I've gotten to the point where peanuts are as sure a crop as corn. I have seed in its twelfth year—grown organically—and it is acclimated to Iowa.

"I've found that plowing under sweet clover ground is best, when the clover is about 18 inches high. Here this would be around May 10th. The soil should be a rich, black, sandy loam and mellow.

"You can plant the peanuts in the shell or out, whichever you prefer. I gummed out an old corn planter butt drop plate to fit a medium-sized peanut and use the full row width. I plow them, using standard equipment, until the peanuts start to show yellow blossoms."

Experts advise not planting peanuts on the same land oftener than once in 3 or 4 years. The rotation should include at least two soil-building crops —one a winter cover crop. Many growers use cow-peas or velvet beans. For winter cover crops, crimson clover, giant red clover, alfalfa, rye and barley are popular.

Planting is generally done between April 10 and May 10. However, no planting should be done until the soil is fairly warm. The crop should be harvested before the vines are killed by frost. To determine when to harvest, check the foliage for a slight yellowing and examine the pods. If the peas are full-grown and the inside of the shells has begun to color and show darkened veins, they can be assumed ready for harvesting.

Controlling weeds is a major factor in peanut production. For best results, start early—many weeds can be killed just before and just after the peanuts come up. In addition, cultivate shallow at all times. Most weeds germinate in the top one-half inch of the soil.

—L. W. Martin

They're Easy in Illinois!

Visitors to our Illinois garden were curious about the "bright green plants with the pretty yellow flowers." We were raising peanuts. One friend who had been raised on a farm in this area told us he remembered "way back" when they raised peanuts, but he'd never seen any as big as ours.

With the arrival of frost off we went to dig peanuts in two rows, each 25 feet long. The plants had "dug in" below the mulch and we noticed numerous immature pods bearing out the fact that you need a far longer growing season than we have here in Central Illinois in order to produce a really bumper crop.

However, the results from our latest effort were most gratifying. The plants were vigorous and healthy, and each was full of good-sized peanuts which we picked from and placed in flats to dry. After two months (the recommended drying time in the *Organic Gardening Encyclopedia*) the nuts were roasted in their shells in the oven.

PEANUT

From our two season's experience in raising peanuts, we have learned that the seeds should be sown at least 3 inches apart since some do not germinate and thinned to one foot.

—Bette Wahlfeldt

Plant Peanuts with Petunias

Last spring we called upon these apparently dissimilar plants to redeem a bad driveway situation created by a recently finished concrete block fence and an intervening 3 feet of bare earth —too unsightly to be left alone for the summer. We decided on the peanuts because we knew their pretty foliage would make good ground cover and we needed the soil-enriching nitrogen produced by the dense root system. For a companion, we picked the rose-tinted Balcony Rose petunia because it is hardy and requires very little attention. In a few weeks we had a pretty 3-foot border bed—the kind gardeners dream about. Here's how we did it.

For a mass effect, we allowed about 2½ feet between the peanut seeds at planting time to provide room for spreading. We followed a hit-and-miss pattern because we wanted definitely to avoid rows of plants set in a formal design.

Before planting, the peanuts were soaked for about an hour in just enough water to cover them. Then, with a small hand tool the ground was worked only at the spot where the seed was planted—at a depth 3 times its size. Next, the entire lot was well watered, and in two or three days the peanut vines came up to show their fan-like leaves and to cover the area with a lacy green foliage.

The peanuts received no other attention except for frequent watering. When the delicate yellow blossoms appeared underneath the vines, they produced sprouts which turned down into the ground and buried themselves to grow into peanuts harvested later in the summer after the vines turned yellow. A large mass of roots formed on each plant to produce clusters of nitrogen-rich nodes which enriched the raw soil.

—Thelma Bell

PEPPER

As a House Plant

When they start fruiting in fall or early winter, varieties of ornamental pepper make very decorative and useful indoor plants. The two best types are the cherry and cone pepper which have small red, purple or cream-colored fruits and green foliage.

Sow the seed in flats about mid-to-late July. Give flats some protection by placing in partial shade such as you get from pines. A light covering of pine needles will prevent soil packing during heavy rains. A cold frame will serve better, if you have one.

When seedlings are large enough to transplant, set them in two-inch pots until "pot bound"; then transplant to 4- or 6-inch containers. Peat pots would be fine for the first transplanting and clay pots for the second.

Leaf Test Shows When to Apply Nitrogen

For the best yields of sweet peppers, a leaf tissue test can show when the plants need nitrogen. By measuring the nitrogen content of matured pepper leaves when the plants begin flowering, research workers have found the yield can be predicted. Optimum production, say scientists, calls for the leaves to contain about 5 percent nitrogen at that point. If additional fertilizer is needed, it can be added early enough to increase the yield.

The testing was conducted by Agricultural Research Service soil experts at the Texas Experiment Station in the pepper-growing Rio Grande Valley. They found that nitrogen amounts in the leaves decreased during the heavy fruiting period, suggesting that the nutrient moved from the leaves to the fruit. While still experimental, the leaf tissue test is expected to be developed into a routine method for pepper producers, and the technique holds promise of improving fertilizer applications and timing for other crops.

Transplanting? Go to the Root of the Problem

Ed Marston is a New England farmer who has been following the dry-root method of setting out his pepper plants.

He pushes the soil aside till he feels some moisture, making a depression about 4 inches deep. Next he lays the bare-root plant down in it at a 30-degree slant, so the rootlets— which grow on two opposite sides of a pepper root—are in contact with the soil. Then he covers the roots and one or two sets of leaves with moist soil, and puts dry soil on top. The roots reach down for moisture, so the plant—usually set out about 9 a.m.— becomes established right away instead of needing several days like those transplanted with very moist soil attached to their roots. And if a grubworm cuts the pepper at the soil line, there's no loss; Ed just uncovers an-

When pepper plants get too bushy, cut the branches out from the center when pruning. Peppers from pruned plants ripen sooner, and there is room for all to grow fully.

other set of leaves. He sets from 24 to 30 pepper plants about 20 inches apart along the single row.

When his plants get too bushy, Marston prunes them from the center, cutting out branches so they're left somewhat bowl-shaped. Peppers from pruned plants ripen at least a week earlier than those left unpruned, and there is room for each ripening fruit to develop properly. Marston's crop starts to ripen around the first of August, each plant producing about two dozen big, thick-walled peppers.

—Devon Reay

Want to Grow Early Peppers?

It is possible and very practical to have early green peppers as well as tomatoes—and what a mouth-watering combination they make! Here is the method we've used for many seasons with invariably excellent results.

About the first week of May, weather permitting, we dig a dozen holes in the garden, two feet apart. The holes are about 8 inches deep. These holes are then filled with composted soil to within 3 inches of the top. The pepper plants (we like California Wonder) are short and sturdy, rarely over 6 inches tall.

The plants are set in the holes at least two inches deep, packing the soil firmly about the roots. Add enough soil to build the hole up level with the surface. Now, place a square of black tar paper over each plant. The squares best suited are 18 inches square, and have a hole five inches in diameter cut out in the center. Make squares as shown by sketch.

The hole permits slipping the square over the pepper plant. Make as many of these squares as you will have plants to set out; since these squares are long-lasting and will serve for many years, it is best to make them from first quality, heavy-duty tar paper.

Next you have a choice of using hot caps or glass bells made from gallon jugs from which the bottoms have been removed. Or you may use both types of covers. Place earth or

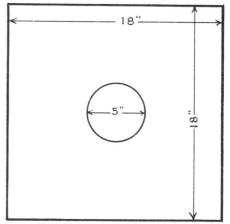

Work with 18-inch squares, and have a 5-inch hole cut out in the paper's center.

120

Set a bottomless jar over the plant on top of the paper, and spread around it a circle of soil so heavy winds can't tip it over.

Use only white glass—colored glass will not encourage plant growth. Tie a heavy cotton cord like mason line around the bottom of the jug. With a fine paint brush thoroughly soak the cord with gasoline. *Keep the gasoline can closed and well away from next step.* Set the cord aflame with a match. The instant all flames die away, dunk the heated portion of the jug into a bucket of cold water. The bottom will fall off cleanly. These jugs may be used later in starting cuttings of various kinds. They may be used over and over during many seasons.

Keep the jugs clean and sparkling so that all possible heat and light may be trapped within them. Dirty jugs will not perform satisfactorily. During dry spells the plants can be easily watered by simply removing the jug and watering the exposed soil. The tar paper blanket will prevent quick evaporation. Best of all there is no weed or pest problem to consume the gardener's time.

—John Krill

stones on the corners of the tar paper squares to keep winds and storms from blowing them away. Place the hotcaps or glass jugs over the plants. Do not leave the jugs corked, keep the jug mouth open.

Here's what happens: The heat of the sun is absorbed by the black tar paper. The soil beneath it is warmed as a result. At night, the tar paper acts as a blanket against the escape of the trapped warmth. The jug or hotcap also traps the sun's heat, thus warming the air within the container, bathing the plant in warm air.

Consequently, with roots imbedded in rich composted soil that is warm from the trapped heat, and with its upper section in warm air because of the hotcap or jug, the plant will grow vigorously. We generally start no more than a dozen by this method. Peppers from these plants will keep us well supplied until those planted later in the season come into bearing. This is also a good way to start tomato plants for early ripening fruit too.

A word on removing bottoms from glass gallon vinegar or bleach jugs.

One Variety is Good—Two Are Better

It didn't take me long to find out—just one season—that while one favorite pepper variety is good, two are better when you grow them as a team and give them organic treatment.

I started them off in flats, 14 by 26 by 6 inches deep, two-thirds filled with equal amounts of organic soil, sharp sand and vermiculite. You can also use a clean can set at the south window because peppers basically aren't hard to grow if covered with 3 to 4 times their own thickness, by a growing mixture that is kept damp but never soggy.

Some folks like to cover the containers with glass or paper until the plants are well sprouted before setting them in the sunlight. But I like to watch them sprout in a warm room, and find I have less trouble with damping-off and subsequent transplanting.

After the small seedlings have set about 4 true leaves, I make rows two inches apart in the new and deeper flat, and begin transplanting carefully. The small plants are put gently into the previously made rows about two inches apart and the soil is pressed carefully around them. After each flat is finished, it is watered well to set the plants a bit, then placed on shelves 4 or 5 inches below our south basement windows at house temperature. The planting of the pepper seeds is done between the first and 15th of March, to have them ready for transferring to the garden around mid-May, our average frost-free date.

To harden off the young plants about the middle of April—they're about 4 to 5 inches tall—I carry them out to my double-sheathed shop which, with two tall windows on the south and two small ones on the west, gets plenty of sunlight. The plants do fine here except in the coldest weather when I cover them—just as you would a cold frame.

Outdoor planting rows are spaced at 3 feet in a pepper patch which gets full sunlight two-thirds of the day. Each plant goes into a foot-deep hole set 30 inches apart, two-thirds filled with compost or manures, and sprinkled with a half-cup of bone meal and a quarter-cup of ground granite. The mixture is tamped down firmly, and the holes filled to ground level with compost or good garden soil.

When the plants are in and watered, I tuck in a 3-inch toothpick or match about half-way down, close and parallel to each stem, and firm them close together. I have yet to find a tomato, pepper or cabbage plant cut down by a cutworm when the stick and stem stay really close.

Next, a 4-foot stake is thrust 12 inches down into each hole on the north side, about two to three inches from each plant. Then, still on the north side, I push a 5- to 10-inch-wide board into the ground to shield the young peppers from the wind, and also to reflect the sun's heat down and back along the planting row.

Combined with the large brown paper sacks we have saved from the grocery store, these stakes and windboards help in covering up the plants when we get a heavy late frost. We cut a small hole in a bottom corner of each sack, and slide it down and over the stake like a tent, anchoring it to the ground with a couple of clods of earth. The wind-board keeps the sack from pressing against the plant, and last year I used my set of sacks 4 times to save my plants. After all danger from frost has passed, I remove the wind-boards and tie each plant loosely to its stake with wide

A densely-leaved pepper plant is supported by a pipe and protected by a wallboard.

cloth strips. Additional ties are made as the plants grow taller, particularly where side branches become heavy with half-formed fruit.

When the pepper plants are near blooming, I side-dress each by lightly sprinkling bone and granite meal around it, working the two thoroughly into the soil with a 4-tined hoe. Next, the whole bedding area is sprinkled lightly with bone meal to add nitrogen, until the sawdust mulching I now use starts to decompose. Last year after side-dressing, I mulched the pepper bed heavily with 4 to 5 inches of sawdust, mounding a 6- to 8-inch circle around each plant to form a watering basin around its stem. It looked a lot neater, worked better, and I have found that peppers need lots of water every week for thick-walled, juicy maturity.

The sawdust between the rows packs down when walked upon, giving a good all-weather walking surface and prevents the growth of weeds more effectively than hay or straw since both have a tendency to reseed themselves. Under the plants, the sawdust remains loose enough all season for the water to work down through. Throughout the season we had an overabundance of Tokyo Bell and Yolo Wonder from about the first of July until a killing frost. We had ample peppers for twice-a-day table use, some for giving, and many with which to can several flavorful relishes and pickles, a winter's supply for ourselves and our families away from home.

Both varieties did equally well to give me a continuous crop of big, meaty, juicy peppers from the first of July until a heavy frost. Tokyo Bell started first, tapering off later both in size and number, while Yolo Wonder came on strong by the end of July, continuing until the end of the season.

These results confirmed my theory that a combination of early and standard plants will, when organically raised, produce succulent peppers for a longer period. By alternating rows —planting next year's peppers in the sawdust-mulched spaces between this year's rows—I hope to use this special bed almost indefinitely—just remove the sawdust, use it elsewhere, and plant.

—Edward P. Morris

Start from Cuttings

I start a few peppers from cuttings taken from the fall garden before it cools down too much. I root these in sand and put them in garden soil to grow over the winter in a cool light place. In the late winter I nip off cuttings, and when spring comes I have a great many little plants ready to set out.

These are my favorites for the garden. They settle into bearing much more quickly, do not succumb to cutworms that endanger the tinier seedlings, and resist a spell of cold weather

Strawberry plant tendrils reach up a pole standing next to sawdust-mulched peppers.

more easily. For myself I'd choose these cutting plants every time, and as selling items they have much to recommend them. They are larger plants with more plentiful root systems that should command higher prices.

Disposable coffee cups come free in any number of eating places. Wash them and prick small holes in the bottom and sides with a fork. The basic facts can be written down on them— variety and date of transplanting. My pansies grow to maturity and bloom in them.

Put a shallow layer of compost in the bottom, then fill with transplanting soil and set the plant in. Turn each plant frequently so that each side faces the sun. I put each plant in its own container for several good reasons. 1—Each plant has the chance to become well developed and rounded. 2—Its root system is not entangled with another plant's, or crowded by it. 3—At setting-out time it is a simple matter to cut away the bottom and the sides of the cup, leaving the roots intact. Cut off the circlet of the top, and slip it around the plant after it is set out, pressing the circle of cardboard into the soil as a foil for cutworms.   —Dorothy L. Baker

POTATO

On Top of the Ground

On Good Friday, I placed cut potatoes on top of the ground and then forked bright straw over them to a depth of from 4 to possibly 24 inches. They were given no more care all summer. When the potatoes planted next to them in the conventional manner had been hoed twice and debugged twice, little green shoots were just beginning to show through the straw. But how those plants did grow after they got started! August 15, they were still in bloom but I turned Pandora and peeked. Right on top of the ground lay big, light colored, perfectly clean potatoes, all ready to be

gathered without the work of digging them. Where the straw had been the thinnest the potatoes were slightly covered with soil. The largest potato weighed three quarters of a pound and lay under the deepest straw. How much larger they might have gotten no one will ever know, for gathering them was so much fun, I didn't stop until they were all picked.

It had been a very wet and cold summer and weeds had almost taken the rest of the garden but only under the lightest covering of straw did any weeds get started. The soil was moist and each particle separate but the overcoat of straw kept the fine grains

Potatoes should be thickly covered with at least 8 to 10 inches of dry leaves or hay.

from being eroded. I spread the now dark and partly decayed straw back.

Next year there will be another top-of-the-ground potato patch but instead of using bright straw, I'll use the old rotted straw from this year's patch closest to the ground and dead and partly decayed grasses and weeds on top, so that the growth of the plants will not be held up until the straw starts to rot as happened this year.

My totally untried theory is that if one were to put a thick overcoat of straw or straw-like material on a patch of ground and leave it there from late summer until the next spring that no further preparation of the ground would be necessary and that seeds of various sorts might be planted under this overcoat, taking off a large part of the mulch for plants that do not grow tall. My idea is that the high mulch has been of benefit to the bacteria in the soil and that plants will grow better in such ground even if none of the mulch is returned but that returning as much mulch as possible will still further help the bacteria and will conserve moisture and protect the roots from the hot sun. Still another benefit is that the roots will be able to stay in the richer top soil to feed.

—Florence Gunn

Also Above Ground

Professor Richard Clemence of Wellesley College in Massachusetts likes the above-ground method too, since he gets a heavy potato yield and can "inspect the vines from both sides occasionally, taking care of a rare potato bug or a bunch of eggs that the ladybugs have missed." Prof. Clemence makes double rows, 14 inches apart on top of the old mulch, with the seed the same distance apart in the rows.

After laying the small whole potatoes in straight rows, he covers the rows with six or eight inches of hay. When the blossoms fall, Clemence begins to harvest new potatoes. To do this early picking, he moves the hay carefully aside, separates some small potatoes an inch or two in diameter from the stems, and gently replaces the hay. (He rates Irish Cobblers best to eat this way.)

Potatoes in "Strips" of Leaves

The first time I tried planting potatoes in leaves, I made the mistake of pulling the leaves back and sticking the potatoes down under them next to the soil. The spring rains were heavy and the leaves kept the potatoes too wet and some rotted. So that is why I now plant on top of the leaves and cover with a good heavy mulch at planting time.

When it is time to dig the potatoes, I just pull the mulch aside and harvest.

The potatoes are the best-tasting, smoothest and largest I have ever grown. I might add that so far I have never seen a *single* potato bug on the potatoes I grow this way.

I have worked out a way to make compost in strips that lets me make two to four inches of humus a year in my garden soil.

Every year I choose a strip of garden from which I have harvested peas or early corn. After the crop is done I simply start piling manure, leaves, sawdust or anything I can accumulate on long strips right over the old vines or stalks. In other words, each row has become a long compost pile right where I want the compost to be—no hauling to be done.

Far more satisfactory than the results from the strips of manure, etc., are the results obtained from strip composting leaves. This, I must tell you, is my "baby," my favorite. In the fall I choose a very wide strip of garden where I want my potato patch to be the next year. Leaves are piled all over this spot almost 3 feet deep.

By spring they have packed down and already the earthworms are hungrily working up through them. When the time is right, I plant the potatoes by laying them in long rows right on top of the leaves. The potatoes are then covered with 12 or 14 inches of straw. Sometimes I have to put a little more on later on if I see any potato tubers sticking through.

—Lois Hebble

—And in Shredded Maple Leaves

We like our potatoes small, tasty and tender; thin-skinned spuds that we put in our stews entire, without peeling or cutting, to simmer in their jackets. So we grow them in 30-foot-long rows of shredded maple leaves *above the ground.*

I gather the leaves when they're most abundant, in late October, and use them first, over the winter, to protect my carrots from alternating freezes and thaws. I spread them with a lavish hand—18 inches deep—and "pin" them in place securely with a light but durable wire cage that extends the full length of the row. Whenever I want some carrots, I shift the cage to one side and dig up a good fistful with the garden fork.

Come spring, the leaves are weathered down to a 9-inch depth and the carrots are about through. In March my seed potatoes arrive and are set out in a quiet, shady place to grow sprouts for the April planting. By the time they have put forth quarter-inch buds, the compost shredder is ready to "plant" potatoes.

On a fine, mild day I trundle the shredder out to where last year's carrot row sits under the mulch of weathered leaves. I start at the top of the row and, moving along slowly, scoop up the leaves and run them through the shredder, laying a carpet of fine leaf particles 4 to 6 inches deep and about 15 or more inches wide.

By the time I am through, 30 to 45 minutes later, I have shredded the entire 30-foot row of leaves on which the potatoes are immediately cut and planted. We then spread sifted wood ashes and rock phosphate by the handful and cover everything with a fluffy blanket of loose hay, 18 to 24 inches deep.

—Maurice Franz

First Time Success with New Potato Method

Last summer, for the first time, we raised our own potatoes in our garden, only because the compost and mulch method suggested an easy way to raise them. We had *started* potatoes before but never *raised* any except for the bugs. Also, what deterred us in previous years was the thought of all the cultivating and dusting usually associated with raising potatoes, but we discovered that neither is necessary.

In April we prepared a patch of garden by turning under the winter rye we usually grow to enrich the soil, and then we waited long enough for this to decompose before planting. We next made furrows with a walking plow, stuck a whole potato in the furrow every 6 inches (cutting weakens the plants), and put a shovelful of compost over each potato before covering with soil. No lime was used in this particular compost since potatoes thrive best in a slightly acid condition. After covering the potatoes with the earth turned up at the side of the furrow we spread a foot of spoiled hay over the whole patch. *That's all the work we did on*

Covered with leafmold and hay, potatoes planted on top of the ground have tender skins, cook well, and have good flavor.

potatoes until time to dig them up.

The potato plants pushed up through the hay, but the weeds did not. Soon the plants were profuse and dark green and kept growing and flowering throughout the worst drought in history in our locality. At no time did we spray. After 40 days of drought, potato bugs ate off many of the leaves, but when it rained, new leaves came out and these were not attacked. We noticed this same curious condition with our lima beans. When the plants had no water for over a month they became unhealthy because of this lack, but when they got a drink they regained their resistance to bugs. Even in spite of the drought, there was no scab or any kind of disease on the potatoes.

Most of the New Jersey potato crop withered up and died in the drought but these potatoes were perfect. Our yield was not as large as it would have been with normal or any rainfall, but each plant had from 24 to 36 potatoes on it and we harvested 200 pounds from two, 50-foot rows.

—Alden Stahr

Cutting Tubers before Planting

I well recall a conversation which I overheard about *twenty-five* years ago between Dad and an old potato fancier concerning how potato yields may be increased, and often doubled by the correct method of cutting the tubers before planting.

Later, I questioned Dad, and he showed me what is meant by "feeder fiber" in a potato tuber. Every eye of the potato is fed by veins which run to the stem end of the tuber. That is the way the tuber develops as it grows, and if undisturbed in cutting, the new plant which develops from the sprout-

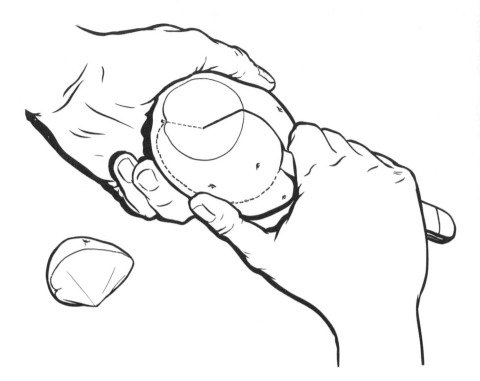

ing eye will develop and feed through the same feeder bundles (or veins) as it fed as it grew as a tuber.

Those of us who like to place yams in a vase of water and have the vines as a winter plant, will notice that the sweet potato feeds in much the same way, until after several months growing, the tuber is almost all consumed except the skin. The Irish potato will feed in the same manner if given a chance.

Now, Dad did not believe in idle talk, so he suggested that the following spring I take over a small plot in a fertile section of soil and cut potatoes in 4 different ways.

In plot Number 1, since the tubers were large, we peeled the potatoes about ⅝ inch thick, cut the peeling to a single eye. In plot Number 2 we cut the potatoes two eyes to a piece and planted them. Plot Number 3, we planted the smaller potatoes out of the lot, whole. In plot Number 4, we cut the tubers, slicing them beginning about ½ inch above the eye, holding stem end downward, actually meeting the stem with the knife in each cut. In this last plot either one or two healthy appearing eyes were left in each piece, which looked like a sharpened post when we came to the eye end of the potato. We then sliced the eye end through the middle to the stem end.

The season was dry, but Dad was liberal with his green and stable manures and his land was in fine condition. When harvest time came the different methods we had employed brought us quite different crops. Plot One where the peelings were planted had grown heavier vines than any of the other plots, but the yield was no larger than good sized marbles. Plot Two, the regular 2-eye-to-the-piece system, had a light crop of potatoes, small but suitable for table use. Plot Three where the whole potatoes were planted was a fine yield, despite the drought, of larger than average tubers and the hills were well filled. Plot Four also had a fine yield. There were no noticeably different results either in size, quality or yield between the planting of whole potatoes and cutting the potato into 5 to 9 sections, so long as the fiber in the tuber was not destroyed.

I have since repeated the methods used in plots 3 and 4 several different years, with similar results. Have finally concluded that for best results, the plant fiber must be preserved both in the land and in the seed.
—C. C. Walker

Treatment after Cutting

When the cutting is finished, I place the potato sets in a bushel basket, then cover them with a slightly dampened burlap sack. These are then stored in our cool basement where temperatures hover around 60 degrees. Each day I go down and stir the sets so they will not stick together. Gradually the cuts suberize, or cork over, thus forming a barrier against insects and rot when planted. When sets are cut and healed ahead of time sprouting is hastened.

It has been discovered by experiment stations that the number of sprouts left to mature on each clump will determine the number and size of potatoes harvested. For instance, one sprout produces a limited number of very large potatoes. Two sprouts give a large number of large potatoes and some medium-sized. Three sprouts give some large, but mostly medium-sized potatoes, with plenty of small ones. If all sprouts are allowed to remain, only small potatoes are produced. I leave two most of the time and receive a great majority of very large tubers.

If you wish, you can control the number of sprouts by sprouting your sets in advance of planting. Place sets in open trays 4 weeks before planting time. The "rose" or that end with the most "eyes" should point upward. Place sets in a light, warm room. Just before planting, remove all sprouts except those you wish to remain. I govern the number of sprouts by waiting until the plants are 6 inches high in the garden. I then go down the rows with pruning shears and snip off all but the two I wish to remain per plant.
—Betty Brinhart

Idahoes in Michigan

The seed potatoes (my own organically grown seed), were planted around June first. The seed pieces were cut several days in advance of planting to allow plenty of time for the cut ends to "heal" and dry properly. Shallow planting holes were made with the point of a garden hoe, and the seed pieces were planted only deep enough to be covered with soil.

After planting, the patch was mulched to a depth of 5 to 7 inches with moldy alfalfa hay. Nature did the rest.

The patch required no cultivation and occasional stray weeds were quickly and easily pulled. Although rainfall during the growing season was greater than usual, it was apparently just right for the potatoes. The vines started slowly, but when once started they grew vigorously, survived a heavy hailstorm, and by midsummer the tops had formed a solid mass of green over the entire patch. In our experience, the Idaho has shown a remarkable resistance to both scab and potato beetles.

They continued to grow and hold their green color until killed by a heavy frost around the middle of September. A couple of weeks later the "fun" of digging began, and the big surprise was really on.

The small garden-size patch (6 rows about 40 feet long) yielded almost 6 bushels of the nicest Idaho baking potatoes one could wish for—a record for our garden in both quality and quantity. Digging was easy. Although our Michigan fall was unusually warm and dry, the potato soil was moist and friable. Many of the potatoes could actually be "picked" as one only needed to move the mulch aside to find them growing practically on top of the soil underneath the mulch.

—Harold Ratcliff

Potatoes Among the Front Lawn Roses

Here's the method I used to plant potatoes among the roses on my front lawn:

I cut the potatoes in fairly large pieces, being careful not to break or disturb the sprouts. I had only 11 pieces available to plant. At one end of the row of roses, the bushes were low and small. So I began planting between these smaller bushes, placing the 11 pieces of potatoes about one foot apart. In order not to disturb the roots of the roses, I used the method of planting on the top of the ground.

Potatoes prosper among low-growing rose bushes when spaced at one-foot intervals.

Each piece was placed on the ground —not in the ground. Each piece was gently covered with a mound of compost, great care being taken to protect the growing sprouts. My small potato patch was now ready for the late spring rains.

In less than two weeks, the plants began to peep through the compost. It was then that it occurred to me potatoes did not really belong in the front yard. First remarks came, of course, from my family. "Mom, why did you plant potatoes in the front yard? They'll spoil the beauty of the roses."

They did seem out of place to me, too, but those potatoes grew so fast and had such a healthy appearance, I just couldn't remove them. So I let them grow along with the roses.

The shade from the potato vines did not harm the rose bushes in the least. Quite the contrary—the rose blooms were larger than ever and, I believe, more colorful. This may have been due to extra fertilizer added at about the second month. Two crops on the same ground should have plenty of fertilizer. Also at intervals during the growth of the potatoes, I added extra compost, sawdust and mulch to prevent sunburn on the new potatoes, as they tended to come to the top of the ground.

Now for the harvest. When the vines had turned brown, I knew it was time to find out what was under those vines. One by one they were lifted from the ground. This was an easy job because the potatoes were on top of the ground. Just pushed aside the mulch and sawdust—and there they were. No shovel or tool of any kind was necessary. The potatoes were smooth and clean. I harvested a total of 45 pounds from the 11 original

pieces. The majority were an ideal size for baking. Some were larger.

—Cora Brehmer

Between the Rows

Most every year in late May, I clean out the remaining potatoes from our kitchen bin and cut them up for seed. This usually amounts to a half peck. Because nearly all of them have sprouted, it is easy to cut them into chunks with one or two "eyes" on each piece. On the average I get 4 or 5 seeds per potato.

These I plant in between the rows and in any open spots of my vegetable garden. I'm not interested in getting a big harvest, but enjoy the fun my family and I have in seeing how these bits of vegetation produce a brood of potatoes in the fall. Their maze of hair-like roots also add lasting benefits to my soil with the organic matter they leave behind.

Naturally I have to use the variety of potatoes that we have on hand— usually from Aroostook County, Maine. You can select seed potatoes from the popular catalogs or at your local garden store if you are interested in planting a larger crop.

I plant my cut-up seeds about 5 inches deep and 3 feet apart, after working a handful of dehydrated cow manure into the soil and stirring it up. Other than an occasional hoeing and another application of manure worked around the plants in July, they take very little of my time. My vines have been surprisingly free of potato bugs, but I have had some in the past. When I see any, I simply pick them off and destroy them.

In the fall, when the plants have died down from frost, the family and I go out to "discover" our harvest. One never knows for sure what will be revealed until a plant is dug up. We use a garden fork to push gently into the soil, and as we lift up a clump of the earth, turn it over, sometimes 3 or 4 good-sized potatoes will appear, their white skins contrasting with the dark soil.

We put them in baskets and take them home to enjoy boiled, mashed or cut raw and pan-fried with onions, until every last one is gone.

By growing potatoes in small lots, there's little work and more incentive to try again next year. I rotate my planting each season by switching them to the other side of the garden. And to avoid encouraging disease, I don't plant them near or in soil that has grown tomatoes the previous year.

—Walter Masson

Spuds in a Compost Box

Since we are short of growing room but have a large backyard patio, we placed a wooden frame, 4 feet square and 6 inches high, on the concrete back patio and filled it with one-third peat moss mixed with two-thirds well-rotted compost. It had been made that previous winter from our autumn gathering of leaves, well-laced with vegetable refuse from the kitchen, and cattle manure. Although tempted, we put in no soil at all. The fluffy heap of organic material averaged 6 inches, rising in the middle to a foot higher than the sides of the box. Into this mass we placed the tubers, and turned our backs on them, not expecting much to happen.

Last spring was unseasonably warm and dry. We feared that our experiment would dry out, but the advantages of mulch proved themselves again. The mulching material, which formed the growing medium, was slightly moist when we first poured it into the frame, and it was 3 weeks before we sprinkled a little water into it. The mulch dried slightly on the surface to become highly effective insulation against the prevailing heat and dryness.

The potatoes in the box continued to surprise us with their rapid growth, forming big healthy vines with dark-green leaves. Although they got the jump on the spuds planted in the ground, once the latter began to come up, they quickly outdistanced those in the box.

We found in later experiments with organic mixtures that about half compost with half peat moss works the best. An additional shovel of steer manure for every two bushels of the mixture also seems to help. I am still experimenting with mixtures and I suspect that a little sand would help the drainage.

I do not claim that potatoes grown by this method are superior to those grown in the earth. Potatoes planted in the earth produce more and bigger tubers, grow larger, and maintain their vigor longer. The skins on the compost-grown potatoes are slightly thicker and appear darker than the ones of the same variety grown in the soil.

What is appealing in the compost-grown potatoes is that they offer an area for experimentation to those organic gardeners who are searching for new and different ways to enjoy gardening. Certainly, potatoes grown in purely organic matter provide one of the more dramatic ways of showing off the effectiveness of organic gardening. For the gardener who is hampered by a small gardening plot, compost-grown potatoes offer just one more way to get the most garden out of the least space. And for those whose chief pleasure comes in eating the produce of their garden, such potatoes provide that same extraordinary delight that home-grown produce always gives.

—John J. Meeker

Hay-Mulching Tests

Our purpose was to determine the relative value of hay versus the different plastic mulches on early potatoes. We planted 9 100-foot rows of Irish Cobblers with hay mulch, and 12 50-foot rows of Red Pontiac with 3 kinds of plastic—green, black and clear.

Spoiled hay was spread 6 to 8 inches deep over 3 rows of seed potatoes immediately after they were planted on top of the ground, *not under it*. Three more rows were planted in the soil, allowed to germinate and reach a sprout height of at least 8 inches before they were mulched with hay. *These were the potatoes that did best of all.* Three more rows of soil-planted cobblers were then left unmulched to serve as controls.

Since potatoes appear to prefer cool, moist soils, we applied the plastic mulches—black, clear and pale green—with great anticipation because the different colors vary greatly in their effect on soil temperature. Clear plastic lets the light rays through right into the soil practically without loss, and then holds the heat inside—just like a greenhouse. It has one great drawback; it admits so many of the sun's rays that weeds flourish inside the plastic undisturbed.

Black plastic soaks up the light energy, converts it to heat, and then passes it along to the soil, but does not permit weeds to grow. Black is a good heat conserver, and gives it up

slowly at night, which makes for uniform soil temperatures over a 24-hour period. The pale-green plastic was the next best potato producer after the hay. We found that early potatoes handle heat well, and did benefit from the ability of plastic to keep things moist in the planting row.

The hay mulch did a fair job of keeping weeds down, while the black and green plastics did very well, as expected. From the sixth week on, the weeds began coming through the hay, and wild morning glory became a really serious problem. Colorado potato beetles attacked all the rows simultaneously, but were much more destructive on the unmulched and plastic-mulched plants, while the hay-mulched crops came through with the least damage.

The Irish Cobblers were the heaviest yielding, but the Red Pontiacs were larger and more uniform. In our opinion, *there is no doubt that potatoes planted in soil and mulched with hay gave the best results.* The plastic mulched spuds were larger than the controls, but their yield was not necessarily greater. Green plastic was the most productive, and the soil under it was much the coolest, although it heated at about the same rate as the bare soil.

—Kenneth Polscer

A Little Girl, but Big Potatoes

Straw mulch was the answer to bigger potatoes for Edna Wirth. The potatoes were solid and crisp and the biggest weighed 2½ pounds. They were grown with natural fertilizer and prospered even though their Illinois country experienced a bad drought. There's proof again that organic gardeners grow the biggest and best produce.

—John Ostertag, Jr.

133

POTATO

My Potato-Planting Secrets

Early each spring, a few weeks before planting 5 pounds of early potatoes in two double rows, we spread old compost generously over the area. The soil has been getting richer every year anyway, for we have buried fruit and vegetable residues along with eggshells in various spots just under the surface for the past 6 years. Just ahead of planting, late in April, we hand-spade only where the rows will lie, to aerate the soil and mix in the compost. A space wide enough for one row of Black Valentine beans is left between the rows.

Next we rake the soil level and lay a long 10-inch-wide board down on the first row. Standing on this, we mark off indentations 14 inches apart on either side of the board in which to place the seed potatoes. (Last year we bought Kiswick potatoes and cut each into 3 or 4 pieces. These should be left to dry for 12 to 24 hours; we forgot, yet noticed no adverse effects.)

After the seed potato pieces are staggered, cut-side down, along both sides, we slide the board ahead to place another section of row. When all potatoes in both double rows are placed, we mark the row end with sticks, then sift loam through a wide-meshed wire basket over the potatoes so each is barely covered. They would probably do as well without the loam over them, since we cover each row with an 8-inch layer of pine needles, which makes a light, airy mulch. Last spring, right after we planted, a week of cold Massachusetts weather set in, followed by a rare spring drought. There was no sign of a potato plant for so long we thought we might have to replant; but it was moist down under the needles, and earthworms were active. All at once the plants broke through, dark-green, vigorous and stocky.

Our potatoes have never been bothered by insects or disease, but then a healthy clump of horseradish always appears, unplanted, midway between the double rows of potatoes. It's a good natural repellent for us.

Besides the plants we sow each spring, stray potato vines come up here and there all over the garden, usually from broken and discarded potatoes or those overlooked at the last harvest. We have even dug small delicious Idahos which must have grown from bruised spots cut out of market potatoes and buried with vegetable trimmings.

—Devon Reay

They Said We Couldn't Do It!

Some friends declared it practically couldn't be done—growing a late crop of potatoes, that is, here in our southeastern section of Oklahoma. Summer's too hot and dry, they said. If you were lucky enough to get them up, the crop would seldom be able to withstand the heat and drought that was sure to come.

Getting late potatoes up and off to a good start was just one of the problems encountered. For when we were successful enough to get a stand and keep them alive, we were almost always in for disappointments at harvest time, with poor yields and potatoes too small to be of much value. Now, however, we've mastered some of the difficulties and would like to pass the information along.

Potatoes planted at any time of year will usually require 3 weeks or more to come up. Late ones will sometimes lie in the ground, straggling along one or two at a time, for a period of 6 or 8 weeks. Therefore—

coupled with lack of moisture and other deficiencies—they're too late to mature a full crop. So we try to get them into the ground the first of July instead of August as had been the habit. Even then, we often have to protect them from the first killing frosts to reap a good harvest.

Back in the fall of 1965 we had built an 8-by-10-by-3-foot compost pile. Into that heap went about everything—sawdust, cow manure, garbage, waste paper. Then, just before the first killing frost, all the green garden plants—sweet potato, bean, tomato, and pea vines and others—were added, together with big cotton sacks full of leaves, supplied by a huge oak at the back of the garden. It all mingled with soil and water to form the very best organic fertilizer.

The following spring we found we had more compost than we could use on our small garden plot, so on the first day of July 1966 my husband spaded up the remainder, measured off rows 18 inches apart across the heap, then filled the little furrows with water from the hose. The next morning, before the sun came up, while the ground was still cool he planted the potatoes about 14 inches apart in the row, pressing large pieces into the soil with his foot. After covering them lightly with earth, he gave the whole bed a 4- to 6-inch mulch of prairie hay, weeds, grass and leaves to hold moisture and keep out the hot sun.

Within 3 weeks those potatoes were coming up with scarcely any stragglers to appear later. At first we had watered the bed sparingly, since too much moisture when first planted may cause the pieces to rot. We learned this by once planting too soon after a midsummer's rain. But now it

was given a good soaking, repeated at weekly intervals thereafter.

Well, those plants grew like weeds in a deserted cow lot. Three feet and more in height they went. One visitor predicted that my husband would get a half bushel of potatoes from that little spot. And he did, 4 of them in fact—two bushels, good big measures, and an extra peck besides. The potatoes were very large ones too, . many weighing more than a pound each. Our surprised friends declared them the biggest they ever saw grown in the fall.

No commercial fertilizer of any sort was used. My husband, C. C. Raney, is 91 years young. He dug those spuds with an old grubbing hoe, and is he proud of them! Needless to say he has another shallow compost heap waiting for this season's planting.

—Ona Raney

Second-Crop Potatoes Are Worthwhile

For the ambitious gardener, a second potato crop can be very worthwhile, according to George Williams, manager of the North Angus Farm in Humansville, Missouri. By using a garden section where earlier crops have been harvested, a late potato yield adds good-keeping spuds to the winter storage.

This is the second year Williams has experimented with late potatoes, reaping an average of 100 bushels to the acre, as compared to the 150 bushel average of the spring-planted crop. His soil is an easy-to-work light sand on the upper edge of the Ozark highland. Normally this area has an April to mid-October growing season.

In preparation for the first crop—planted April 1—Williams turned under a 4-inch manure-straw covering.

This warmed the cold soil and helped fertilize the plants. A late frost nipped the vines a bit, but did no real damage. Rains came at the right time, causing the vines to grow at a rapid speed, and they were practically insect-free. What few bugs appeared were picked off by hand. This crop was harvested July 18th.

Using select, early potatoes for seed, Williams planted his late crop July 22nd, depending on the spring covering for fertilizer. Since the ground was fairly dry and no rain was promised in the forecast, he decided to speed up germination by wetting the eight rows of 100 feet each with the garden hose. As he had hoped, the plants came up quickly and evenly.

"Actually, late potatoes aren't anything new to me," George said. "During my boyhood, most of the farmers planted their patch back, shortly after summer digging. But I can't remember ever having seen anything but a poor stand and a lot of spindly vines that yielded no more than one small potato. I wan't interested in this sort of thing."

Williams believes the secret is using lots of barnyard manure—plentiful at the farm's stock-hay operation—and making certain the ground is wet down thoroughly after planting. "This method has given me vines that were as tall and verdant as any spring crop," he said. "And I wasn't bothered with insects at all."

The day following the first frost of the season—October 18th—Williams dug his potatoes. And although his spring crop is beginning to shrink and wrinkle, even though he has an excellent cellar, the fall crop will be crisp and tasty well into summer.

—Marie Walston

Leaf Covering Brings Bug-Free Potatoes

Two years ago I did not dig up all of my potatoes, and the following year several "volunteers" appeared. These potatoes were the largest and best I have ever grown.

When I uprooted my crop last fall I buried the small ones and covered them with leaves. In the spring they started to show through the layer of leaves. All this time I did not touch them, but let them grow "wild." I gathered this last crop in September.

In the past, I sprayed my potatoes numerous times because of bugs, but this year, due to the leaves, there were no insects bothering the plants.

After gathering the year's crop, I spaded under the partly decayed leaves and planted the potatoes that were too small for table use. Then 6 inches of dirt was added on top of the leaves and straw was spread over the ground. With these precautions, the plants will not have to be touched until time to dig them up.

—Cecil N. Newsome

Plant Potatoes with Onions

Planting onions right with the spuds has been a great help to us. We not only got fine onions with the potatoes in the same plot of soil, but the onions definitely discouraged insects feeding on the vines, and also others feeding beneath the soil. I can't explain exactly what effect the onions have on the pests, but I have witnessed over the years that small control plantings without onions harbor more insects than those combined with the pungent bulbs.

We continue to use this method with potatoes. It is interesting to see what happens when onion stems are crushed near some plant crawling with insects. The heavy onion odor saturates the air, and the pests depart. I believe that the volatile oils released by the crushed green onions coat everything with a flavor that is highly unpalatable to the pests. —John Krill

PUMPKIN

Mom's Rampaging Pumpkin Vine

Our farm has never seen chemicals. For 15 years we carried the kitchen garbage to the farm while we commuted from the city. It was carefully composted.

We have big trees at home and we were not able to use all the leaves there, so the excess has also been carried to the farm for compost, even though there are enormous trees at the farm. Wood chips from trucks cutting and trimming trees have been composted—and, in short, the farm has never seen chemicals. The result has been that roses, vegetables and perhaps a few weeds have grown as though that were all they had to do.

This year Mother put a pumpkin seed in a spot where the last kitchen garbage had been composted and allowed to rest for a year. This pumpkin came up like Jack's beanstalk, but instead of growing vertically it grew horizontally—as all good pumpkins do—and it grew errantly as though it were my sister's children toddling in all directions at once. It headed for the woods and then, as though knowing it would not find any sun there, turned 90 degrees at the very edge and headed for the wood shed. It went through the vegetable garden, past the grape arbor, and had decided to head for the pear tree.

We used some of the young pumpkins as "squash" and still got 14 nice big healthy pumpkins this year. The marvelous thing about it is this rampaging pumpkin vine did all this without any assistance whatsoever from the chemical industry or the chemical lobby in Washington—can you believe it? It had nothing but the rotted kitchen garbage which many people's expensive disposals wash down the drain.

—Mrs. D. H. Payne, Jr.

Fit for a Cinderella Coach

Out in Salem, Oregon, the Kelley Peters' family discovered their volunteer pumpkin measured 7 feet around and weighed 161 pounds. Their 6-year-old boy claims possession, while

137

PUMPKIN

Mrs. Peters credits the "black-sheep" compost idea as one factor in the soil's fertility. Another OGF idea—interplanting for pest control—also helped, she reports. "I planted herbs amongst the vegetables, as well as marigolds and garlic, and had very good luck with them as insect repellents."

Another pumpkin, this one grown by Otto Barbush, Sr., of Weatherly, Pennsylvania, led to some interesting developments. Barbush planted the seed about 5 inches deep, surrounding it with several handfuls of a mixture of humus, earth and cow manure. Over each planting he put a mulch of old leaves, straw and well-aged horse manure. With some dry-weather-watering aid by young John Postupack and his brother Jim—on whose parents' farm the planting had been made — the biggest Barbush pumpkin attained a height of 23 inches and weighed out at 129 pounds. After cutting off a large piece from which his wife made 10 "tasty and smooth" pumpkin pies, Barbush donated it to the Carbon County Home, where another 104 pies were made — a total of 114 pies from one organic pumpkin!
—M. C. Goldman

Plant Pumpkins to Foil the Frost

As the pumpkin is not frost-proof, we plant in May here in Arkansas, well after the last freeze, and continue planting until early August. However, since they need about 120 good, hot growing days, we advise the gardener who is trying pumpkins for the first time to stay well within the frost dates for his area.

In the northern states where the growing season is short, plant pumpkin seed in individual pots in the hotbed or greenhouse, transferring them outdoors when the soil has warmed up.

For the garden, sow seeds two to 4 weeks after the last frost. In rows, sow one or two seeds to the foot, later thinning to two to three feet apart. Plant two or three seeds to each hill, thinning to a single plant, spacing the hills 4 to 8 feet apart. The seeds should be planted one inch deep.

Soil should be loosened and raked around the stalk vine to encourage secondary root development. Compost and mulches may be added as the stalk grows. When the fruit appears, be sure to set protective layers of plastic cloth or paper under each to prevent insect attacks and other soil damage.

After the stalk vine has grown several feet and is well anchored, it may attempt to climb. It can be discouraged by pruning or clipping, or it may be trained to a tree without fear of damage. We have had several tree-trained vines produce 15- to 20-pound pumpkins and often have a few plants near a bush or a tree. We also have grown summer pumpkins between rows of corn to shade them from the direct rays of the sun.

The roots of spring and early-summer pumpkins go deeper into the soil if not watered and forced along by too much artificial watering. If the leaves begin to wilt in midday, though, they either do not have enough moisture or the garden "stink" bug is sucking. The best way to rid the pumpkin of that sucking menace is to water all along the stalk. The bugs don't like the water, so they will crawl out in view where they are easily caught and smashed. Also, look on the leaves to find the patches of brown-colored eggs. With the fingernail, gently scrape them from the leaf into a container.

Pumpkins planted in early August or July may need some artificial watering to produce before frost. Much mulch and some shade are also very helpful at this time of the year. It is especially useful if the stalk develops several roots as it grows.

The thick shell covering helps to preserve a mature pumpkin for many months of time if stored in a cool room, at about 50 or 60 degrees without too much light. A cellar, if not too damp, is an ideal place to store pumpkins if insects are not prevalent. Even without air conditioning, we have kept some of the thicker shelled, mature pumpkins from the summer harvest until May of the next year. We've found that they keep better in a northern room away from too much light.

One of our favorite ways of serving pumpkins is in mixed vegetable soups, casseroles and such. Also, baked pumpkin can be a real taste-treat.

—Mayme Bobbitt

RHUBARB

Getting Better Results

After having indifferent results with rhubarb for a few years, we decided to transplant the roots once more (the third time) and to do it properly. We dug them up, trimmed them and kept only the sound, healthy parts. A hole the size of a half-bushel basket was filled with a mixture of compost, organic plant food and topsoil. The roots were planted about 4 feet apart, covered with two inches of rotted wood chips. If this had been done in the first place, we might have saved ourselves a lot of work.

The only further attention necessary is a generous feeding each spring and fall. In the spring, the previous year's mulch is worked into the soil along with the plant food. In the fall, the mulch is pulled back and the plant food worked in lightly around the roots. If the plants have been very productive in the spring, another feeding may be worked in lightly in midsummer.

Early in spring, when this region (southwestern Ohio) has the most rain, the rhubarb shoots grow like magic. When the stems are at least a foot long they are pulled, not cut, from the plant. A good, firm grip on the stem down in the heart of the plant, a little twist, and the stem separates cleanly from the root. The smaller stems should be left on the plant in order to feed the roots during the summer.

—Lucille Eisman

When tall enough, rhubarb should be mulched with strawy manure, wheat straw or hay.

Rhubarb Before Robins

Soil between the plant stools and on each side of the row of plants should be dug early in the spring to kill any weeds that survived the winter. When digging, turn each shovelful upside down so that the weeds will be smothered. Then cover the soil between the stools with a thick layer of partly-rotted manure. The crown of the stools may be covered with a strawy portion of the manure to protect them from the cold.

Very early crops of rhubarb can be harvested by lifting the stools and replanting them in a cold frame. Or, you can place a bottomless bucket or jug over each stool, and then pack horse manure around the outside of it to keep away the cold. Horse manure produces heat as it rots and, as a result, produces the earlier crop. But straw alone can bring a crop a few weeks before the ordinary outdoor one. Keep the opening at the top of the bucket covered with a piece of glass so that heat does not escape. As soon as the regular crop is ready for pulling, remove the bucket or frame, and do no more pulling from the forced stools. Select other rhubarb stools for forcing the following year.
— Peter Keegan

In Time for Christmas

Rhubarb can be grown in the basement this year, in time for Christmas providing you've got a good supply of outdoor growing crowns two to three years old.

You'll have to cover the windows because stalks produced in the dark are more flavorful and succulent than those grown in light.

The planter box is a good place for the rhubarb, but the redwood tub— one crown to the tub—is better. The crowns should be taken from the field after summering for two years. The best kind are those which haven't been harvested, but which have been allowed to grow and develop naturally. Older or younger crowns will not produce as well indoors.

Crowns ready for digging have been in the field for at least 6 weeks after their tops have died and have been exposed to near-freezing temperatures. Dig deeply under the crowns, getting all of the storage roots with the least possible injury.

Make sure each crown has one to one and one-half square feet of space in the tub. They'll need more water in the tub, but are a lot easier to move around.

Since rhubarb is a heavy feeder and prospers on organic nutrients, the best growing mixture is made up of 5 parts compost, 1 part well-rotted manure and 5 parts sandy loam. If it's hard to get sandy loam, try growing your rhubarb in 20 parts compost and 1 part rotted manure.

After planting, apply or expose to mild heat. Allow some of the crowns to rest dormant so you will have them available later in the winter. The others should be exposed to temperatures as close to 60 degrees Fahrenheit as possible—the ideal growing environment. If the temperature is lower, growth is slowed, and if it goes too high, the stalks lose their red color. At 65 degrees, the yield begins to decrease.

Keep the soil moist to maintain sound growth without loss of quality. I have found that the number of stalks falls off sharply when the plants become too dry.

Harvesting should start about one month after beginning heat exposure—just in time for Christmas. The stalks are tenderest at 16 to 20 inches.

How long each crown will continue to give you fresh stalks depends on many factors, including the vigor of the crown and the uniformity of temperature in the growing area. Usually, with good management, harvesting will be extended over two to three months.

If you've followed directions and used sound crowns, each should reward you with about two and one-half pounds of the firmest, tangiest, reddest rhubarb you've ever enjoyed —still another organic gift to be savored with all the good things on Christmas Day.

—Komo Mayamo

"Stalking Technique"

The first spring in our new home, we found clumps of rhubarb growing everywhere—in the thickets that later became our yard, in the orchard, around the garden, and in some of the fields. They were the surviving remnants of several old commercial patches, still hardy in the midst of weeds, grass and briars.

Deciding to give the rhubarb business a try, we began to dig the old crowns, break them apart, and reset in definite plots. This continued each spring—until we finally awoke to the fact that we had 3 patches of rhubarb totaling ¾ acre, plus two rows as a border around the yard! We were literally surrounded with *Rheum rhaponticum*, and though Indiana is a long way from western Asia, native home of the plant, you'd never have known it from looking at our flourishing stands. Yet all we had done was put the crowns into the ground. They had really grown.

All that rhubarb requires is a well-drained soil, light or heavy, but rich in organic matter. Keeping the weeds out helps, of course, although once started, rhubarb will survive for years even when choked with weeds and grass.

Mulch the whole patch heavily during the winter dormant period. When spring comes, the plants will push up through that mulch and provide some of the most tender, succulent stalks anyone could wish for! True, the mulch will slow sprouting of the new crop a little, but that is a small price to pay for being freed from the chore of cultivation. We've used wheat straw more than anything else, but other materials such as hay, leaves and sawdust will also work quite well. Since rhubarb grows best in a slightly-acid soil, mulches like sawdust and leaves have a doubly beneficial effect.

Rhubarb setting is either a spring or fall event—with spring being preferable in most localities because of increased rainfall. For fall setting, wait until cold weather kills the tops. For spring—the earlier done, the better. The best time is right on the heels of retreating winter. Most people who spring-plant make the mistake of waiting too late, and as a result must contend

New sprouts should be set in the prepared trench no deeper than two or three inches below the surface. Spread rich topsoil or compost mixed with soil around cuttings.

with plants that have made too much growth for successful transplanting, as well as weather that has become too warm. The secret is to set plants just as they are beginning to awaken from their winter dormancy. Otherwise, wait until fall or the next spring. As a guide, try not to set plants with new stalks any taller than 6 inches. The smaller the new shoot, the better.

—John McMahan

Rhubarb by the Barrelful

As soon as the sprouts of rhubarb appear in your garden this spring, work some well-rotted manure into the soil around the clumps. Then cover them with a barrel two or three feet high from which the bottom has been removed. Pack soil and manure around the outside of the barrel to encourage warmth and let this old-fashioned method of forcing produce succulent stalks weeks in advance of the normal season.

In the event that a barrel is not handy, a peach basket or a box with bottoms removed will substitute. So will a deep cold frame, if it can be put right over the clumps.

Whatever method you choose for forcing, you can enjoy treats from this ancient pieplant with extra-early servings of its savory sauce, in pies, tarts and other ways. Some folks say that nature gives a rhubarb harvest in the spring as a spruce-up tonic to benefit our bodies after the winter months.

If you have a well-established bed of rhubarb that is getting crowded, divide the roots this spring when the soil is workable, allowing one eye to a clump. Or if you've never grown this herb in your garden before, you can order the roots from a mail-order house or nursery for planting when the soil is workable.

To prepare a bed, dig a trench about one foot wide and up to 2½ feet deep, saving the topsoil and discarding the subsoil as you work. Replace the subsoil with a mixture of compost and well-rotted manure to within a foot below the surface. Then replace topsoil, planting the roots 4 inches deep and 4 to 5 feet apart. Feed generously with well-rotted manure.

Stalks of a new bed should be allowed to grow freely the first season and removed sparingly the second. After that they can be enjoyed for years to come.

If you are not interested in the seeds, the flower stems can be removed as they appear. This will give mo vitreality to the plants. Once established, clumps will require little care except for generous feedings of compost and well-rotted manure.

—Walter Masson

RUTABAGA

A Winter Storage Favorite

Wherever two or more gardeners gather, conversation turns to talk of corn, tomatoes, strawberries, etc., but you seldom hear heated discussions about the cultivation of rutabagas. That's a shame because there is a lot to be said for my favorite winter-storage vegetable. Perhaps, though, people would rather eat them than talk about them.

It's also unfortunate that too many people judge the rutabaga by the supermarket sealed-in-wax type of product which is often bitter and strong-tasting. Once you have grown your own and stored them successfully, you will never be satisfied with this poor-tasting substitute for the real thing. Higher in food value than turnips, this vegetable is an excellent source of vitamin C and a good source of calcium all wrapped up in a succulent and appealing flavor.

Sometimes called the Swede turnip, rutabaga is a partial-season crop and a fairly quick grower that should find a spot in even the smallest garden. After an early-season crop such as peas, or lettuce, you can plant your rutabaga and be assured of a winter vegetable par excellence. In our section of New York (northwest) we plant around the first two weeks of July. It is a hardy, cool-weather crop that will grow in any friable soil.

I simply plant my seeds where they are to grow at a depth of about one-fourth of an inch in rows two feet apart. Seeds germinate in about a week, and when they are about an inch high I thin them to a foot apart. Being a root vegetable, rutabagas do best in a light soil, well worked and with sufficient moisture to make good, speedy growth. They don't require a rich soil, so they can follow other crops without any additional fertilizer. Planting late in the summer has them growing in brisk, cool nights that help them taste all the sweeter.

Commercially raised rutabagas are planted earlier than home-grown, usually around July first. That is done to allow them to reach a maximum size for marketing, as they are sold by the pound. For your own table use, sow later and harvest a smaller vegetable—but with much more quality and flavor in firm-textured, sweet roots.

The best time to harvest your rutabaga is after there has been a light frost, but before the ground freezes hard. I dig or pull up my crop when the ground is fairly dry. After looking them over carefully and discarding any with flaws, I cut the tops off one inch or so from the crown. Next I layer them in crates or bushel baskets with straw, and put them in our barn for storage. They can be kept in a cellar store room in sand, but have a tendency to give off strong odors that might seep through the house. Maximum storage period is from 3 to 4 months. According to tests run at the Montana Agricultural Experiment Station, winter storage has no marked effect on the vitamin B or C content or its potency.

As to varieties, our favorite is the Macomber, a white-fleshed rutabaga. It is so sweet, mild-flavored and fine-textured that once you have tried growing it you probably won't go back to the regular yellow-meat type. If you still like the yellow rutabaga better, American Purple Top has long been popular. This old-timer also makes an excellent winter feed for sheep and other livestock. If you raise them to feed stock, they should be planted early enough, usually from June 15 to July 1, to allow them to attain a maximum size. Other yellow varieties that are popular are Golden Neckless, Perfection Swede and Golden Heart.

—Joan Lindeman

Why Not Sow a Row?

For rutabagas the secret of flavor and texture is rapid growth. They prefer a slightly acid soil—pH between 5.5 and 7. Avoid planting where another member of the cabbage family or radishes have recently been grown, since these heavy feeders use up nutrients rutabagas need most for rapid growth. Golden Neckless, Perfection Swede and Improved Purple Top (our favorite) are all excellent varieties for the home garden.

Plant rutabagas in northern sectors between July 1 and 15. Exact planting dates depend on when your first frost is expected. Since the crop matures in approximately 90 days, that much time is needed to let it grow properly before cold weather sets in. Our first frost occurs during the last week in September here in Massachusetts, so we must have seed in the ground by the first of July.

Prepare the plot for planting by turning under a good supply of rotted manure or compost. If it's been well fertilized for spring vegetables such as peas or lettuce, no additional fertilizer will be needed.

Rutabaga seeds are quite small and hard to handle. Mix with a little sand before sowing to keep them planted thinly. Sow in drills spaced 22 inches apart. (An ounce of seed will plant a 400-foot row.) Cover seed with ¼ to 1 inch of soil, depending on how dry the weather is in midsummer. If soil is dry down to 6 inches, irrigate first, then sow as soon as it dries out enough to be worked. Next cover seed with ¾ inch of damp soil and firm lightly. A narrow 1-inch-deep strip of freshly-cut grass clippings over the planted drill keeps the soil cool and moist, encouraging good germination. Remove this mulch at the first sign of green.

Like carrots, rutabagas will not form large, round roots if crowded. When seedlings are 4 to 6 inches tall, thin to stand 10 inches apart in the rows. When the roots begin to swell, more thinning may be necessary to keep them from touching.

After our thinned-out plants gain some growth, we hill the rows slightly, then break up the crust in the aisles with a grub hoe. Next we mulch heavily with any organic matter on hand—sometimes strawberry plants from the old bed or just dried grass clippings.

Because their roots reach far down into the subsoil (as much as 3 feet within 40 days) rutabagas seldom suffer from mild droughts. However, they can be injured by severe dry spells, so give them a deep soaking once every third week until the rains return.

Don't wait until the 90 days are up to start enjoying your rutabagas. Although they won't have the very mild flavor acquired after a light frost, they're still delicious and can be used as soon as roots reach reasonable size. For winter storage, harvest after the first heavy frost. Let them dry on the surface for a while, then prepare by cutting off the tops and all roots. Store either in a root cellar, in outdoor pits in milder climates, or in boxes of moist sand kept where temperatures remain around 32 to 40 degrees. Do not let them freeze, however, as this causes spoiling. —Betty Brinhart

SALSIFY

Hints for Salsify

Like most root vegetables, salsify is a long-season crop. Since it takes about 120 days to mature, seed must be planted as early in spring as the ground can be worked. In areas where the soil cannot be turned early enough in spring, the plot may be worked in late fall, and the seed planted just before the ground freezes.

After the rains have settled the soil for planting, we stretch a line, make a shallow trench with a hoe, then plant the seeds carefully. Our soil is light, so we must firm it over the planted row to hasten germination. If your soil is heavy, merely cover the seed with about ½ inch of soil. One ounce of seed will plant a 100-foot row; plant rows 15 to 30 inches apart, depending upon method of cultivation.

Once sprouted, the seedlings gain height rapidly and must be carefully thinned. As soon as plants are 3 inches tall, I thin them to stand 4 inches apart. Two weeks later I thin again if they have become crowded. As in the growing of carrots, thinning is the key to thick fleshy roots, so don't be afraid to pull out the excess plants. The more room the individual plant has, the larger, straighter and better-textured its root will be. You will get just as much from 3 plants growing in one foot of row space as you will from 6 plants squeezed into the same area.

Other than scraping the surface to control weeds, the only time I cultivate the rows is immediately after the last thinning. At this time, I work each aisle deeply with the grub hoe to loosen the soil for good air circulation. Each aisle is then heavily mulched with old hay or dried grass clippings.

Salisfy likes a deep, organically-enriched soil that is not too heavy, and that is free from lumps of hard soil or stones that could deform the long taproot. After experimenting with it in our own garden, we found that it does best in a deep sandy loam rich in humus, which we supply by adding decomposed manure or compost and rock fertilizers, such as ground rock phosphate and potash rock. Also, finely ground limestone is worked into the upper two inches of the plot whenever a pH test shows the soil fairly acid, since this vegetable prefers a neutral to slightly alkaline environment.

Rows of salsify remain a crisp green long after the rest of the garden has been cut down by frost and cleaned up for the winter. Until the ground freezes, we dig the roots as we need them, and bury the discarded, onion-like leaves right there in the rows as a soil-improving green manure.

One important thing to remember about salsify is that it likes to continue growing at an even pace. If growth is retarded by drought, the flavor and texture of the roots suffer. We pay more attention to the soil beneath the mulch in the aisles than to the weather itself. If the soil under the mulch feels the least bit dry, we water until it is well soaked down to at least 8 inches. Because of the heavy mulch and frequent summer rains, we haven't had to water more than two or three times a season.

By early winter almost half of the supply has been used. Before the hard freeze sets in, two-thirds of the remaining plants are pulled up. Their tops are cut back almost to the crown; then they are stored in an outdoor pit between layers of straw, and covered with at least two feet of alternate layers of earth and straw. Once a week throughout winter we uncover the pit and remove as much salsify as we need. The remaining third is left in the garden row until the spring thaw, after which they are pulled and stored in the pit where they remain in good condition until well into summer.

Our favorite variety is Mammoth Sandwich Island, which matures in 120 days. Its roots are white, or slightly yellow, and reach 8 inches or more in length with a diameter of 1 to 1½ inches at the crown. This is about the best variety for the home garden. Additional treats you might like to try are the Long Black or Scorzonera and the Spanish Oyster Plant, called the Golden Thistle. Although not true salsifies, they are handled in exactly the same manner, and are about as delicious.

—Betty Brinhart

SEA KALE

Sun, Soil and Seaweed

Sea kale likes good light loam, plenty of sun, no shadows and, where it is procurable, seaweed as its favorite food. Well-rotted manure should be used wherever possible; fish meal, poultry manure or bone meal are effective fertilizers, especially in sandy soil.

To get good shoots in the spring, plant the crowns out in late autumn during a warm, dry spell, about 18 inches apart each way and two inches below the surface. Cover the bed with a good layer of clean, dry leaves, hay or straw and spread a layer of earth over this to keep it from being blown off. Make sure the bed is well covered so that frost cannot reach the plants until spring comes, when you will see cracks appearing. Underneath is your sea kale with an excellent flavor. Uncover carefully and cut each blanched sprout when 6 to 9 inches long. The purple-tipped varieties do not blanch as easily as the white kinds and the flavor is not as delicate.

For winter use, in late fall choose the strongest crowns and put them in deep pots or boxes so that they are 10 inches in the soil and two inches under the surface. Cover with another pot or box and place in the dark, under the staging in a warm greenhouse, window box or hotbed. Give plenty of water and in 5 or 6 weeks your sea kale will be ready for cutting.

When the outdoor bed has been cropped, leave it covered until there is no danger of frost, then remove litter entirely. The plants will produce green shoots; when these are 4 to 6 inches long they can be cut and served, either cooked or raw in salads. Sea kale is one of the easiest of the mineral-rich leafy vegetables to digest.

To propagate, prepare thongs from the clean, straight side roots which grow out from the main root. Those which are as thick as a pencil or thicker should be selected and cut into pieces 6 inches long, the thickest end of the thong being cut level and the thin end cut slanting. This way the gardener will know which is the top. They can be tied into bundles and put in layers of damp sand until planting time and when uncovered the top end will have produced several eyes; all but two of these should be rubbed off.

At the end of March, plant out in rows 18 inches apart, setting the top of the thong an inch below the surface. During the summer, any flowering stems should be removed. When the tops can be seen through the ridge, they are ready for cutting. Use a sharp knife or spade and cut half an inch below the crown. When all have been cut, rake down the soil, leaving stems covered over an inch. Manure the rows well each year and sea kale may be grown without disturbing the rows for at least 6 years.

—Rosa James

SOYBEAN

Landscaping with Soybeans

We planted our bare, sandy lot in suburban Minneapolis from scratch. It was bare, that is, except for a dense mat of quack grass and other weeds. The first spring we put in a number of fruit trees and shrubs, some shade trees, a patch of grass, and the beginning of a garden.

Around each small tree we left a good basin to soak the roots periodically. And around the edge of each basin we planted a ring of soybeans. By the hot part of the summer, bushy rich-green soybean plants shaded the root area of the trees and gave it a little shelter from the drying wind, saving as much soil moisture as they used for themselves. They also served as a protective circle for the little twig against trampling. Toward fall, each young tree was briefly circled by a ring of gold as the beans ripened.

We continued to dig the stubborn quack sod through the summer, and after planting time for most crops was past, we filled the cleared spaces with closely-sown soybeans. They germinate quickly and soon cover the bare areas. Nitrogen-fixing bacteria form nodules on the roots. These remain to enrich the soil after frost kills the plants. Leaves also fall among the small stems, and by spring are decayed enough to add more humus to the soil. We always use a bacteria inoculant on our bean and other legume seeds to speed growth of the root nodules. This is particularly important on ground where legumes have not been growing recently, in which the bacteria may not be naturally present.

We've used quite a bit of White Dutch clover as ground cover, too. Like the beans, it fills the soil with nitrogen-laden nodules. In addition, it blooms freely and fills the air with its sweet scent, attracting every bee in the neighborhood. But clover takes longer and is harder to get rid of once established. Where we want low cover for a longer time, clover is fine, but for a quick coverup or taller growth, we've found soybeans, really fill the bill. They don't need pampering and have hosted no pests.

In an S-curve along the street side of our wisp of front lawn, we started a sweep of slow-growing shrubs. Outside that—for protection from short-cutting children and older trespassers during the few years it would take them to become established — we planted an S-curved hedge of Chinese elm seedlings gleaned from our ground. But the elms too were practically invisible to careless eyes, that first year, and didn't do the intended job. So, on the street side of them, in soil that would grow practically nothing else — you guessed it! — an S-curve of soybeans. They formed a neat, bushy hedge for the summer. Our elm hedge was saved and took over on its own the following year.

A "fence" of soybeans is also handy where a woody hedge would trap unwanted snow in the winter, such as along a driveway. And they can fill in as a temporary foundation planting, with flowers either interspersed or in the foreground. Other plants might do the same job, but soybeans are cheap, about the easiest to grow, and fill the space until you can get your soil developed, decide what you really want permanently, and scrape together the cash to pay for it!

Our vegetables mingle with ornamentals around the yard. In one corner of a flower area we planted an arc of soybeans to frame a niche for 4 unstaked tomato plants. The tomatoes rambled prettily about the space, and then toward fall seemed to snuggle against the beans for an extra bit of protection.

We plant concentric circles of soybeans around our cucurbits, cutting out the inner circles as the vines spread outward—again for soil building and ground cover quickly and briefly. Because of a strong west wind here much of the time, we leave just a few bean plants standing. The cucumber vines cling to them and are kept from twisting and becoming damaged by the winds. Vines hold themselves a little off the ground that way, too, so the cucumbers stay clean and well-formed. When other cucumbers were dying from heat and other midsummer problems, ours were still going strong under the soybean leaves. Don't plant your beans too early if you are going to leave them among your cuke vines. Until hot weather, the young vines need a lot of sun.   —Marraine Miller

Soybeans In Your Garden

Here in Arkansas, I open shallow furrows late in May in freshly plowed ground that has had an additional application of manure and straw bedding from the barn. I hand-plant the Lincoln variety, a round yellow soybean, sowing it thickly one inch apart in furrows of dark, mellow loam spaced widely enough for my 26-inch rotary tiller. As the ground is warm and moist, I cover the seed with about one inch of soil, firming it carefully by walking back and forth down the rows.

After a few days, when the sprouting beans are literally uncapping the earth along the rows, it is time for the first cultivation, and I give the middles a good stirring, tilling as close to the young plants as is safe. About 10 days later, the soybeans are high enough to permit the second step in the experiment—sowing more soybean seed by broadcasting them in the middles—one seed to about each 4 inches, scattering 3 handfuls of seed at every step.

Finally, I run the tiller through the row middles to cover the seed and give the growing plants in the rows their second cultivation. The second sowing, in the middles, is not cultivated after planting, but within a week it is putting forth shoots while the older beans in the rows are now shading them to some extent.

The soybeans grow as the weeks pass until they are waist-high and begin to blossom heavily. Soon pods form up and down the stalks, gradually filling with maturing beans. Finally the pods turn brown, mature until rather dry, and rattle when shaken as the leaves fall from the stalks—naturally. The second plantings in the middle mature only a few days later than those in the rows, and the entire crop is ready for harvesting.

On a midsummer morning, after the dew is dried, I begin to cut the stalks with a sickle-bar mower. If no rain is expected, I allow the crop to dry on the ground for several days, then tie them into small bundles and haul

Use the rotary tiller with the furrowing attachment to make soybean planting rows.

The furrows are made deeper so the seeds are closer to moisture during dry spells.

Till between the rows before broadcasting seed in the middles, one to each 4 inches.

The shredder is used to thresh soybeans, cane, okra seed and pea pods from stalks.

them to the hayloft where they dry for a few weeks. Tying the vines in small bundles makes it much easier to load, unload and thresh the crop.

I use my compost shredder to thresh beans, cane seed, okra seed pods and peas, holding the small bundles of stalks in the hopper until the shredder knives cut the pods off the stems. You can also separate the stalks and run a few through at one time. I prefer my system which allows me to re-shred the stalks and use them for mulch.

To remove the chaff from the seed, pour them through a wide wire mesh to screen out the coarse debris. The remaining chaff may be fanned out when the wind is blowing by pouring the beans from a bucket into a tub.

The Bansei edible soybean is a good table variety, but its pods do not all mature as do the Lincoln and other "field" varieties which are more prolific and a surer crop. —David Criner

SPINACH

Success in Spite of the Heat

Our soil here in Columbia, Maine, is heavy clay, slow to dry out in the spring. Abundant rain is our lot during March, April and sometimes early May. The resulting wet, soggy gardens often do not dry out enough to be tilled until late May or early June. By July the hot summer sun pours down upon the soil, baking it hard.

Spinach, all the gardening publications told us, requires cool weather, fertile soil and plenty of water. So for a few years, we planted in May or earlier, but each time the summer heat produced a few small leaves and just tall seed stalks.

In the summer of 1958, we got interested in mulching and put huge amounts of hay on a 10- by 30-foot untilled pile of ground covered with weeds and grass.

The following April, we removed the mulch and planted 8 rows of Dark Green Bloomsdale.

We cultivated twice during the next two months and watered our plants whenever the soil seemed to be drying out.

On July 8 we enjoyed the most delicious spinach we had ever eaten. Our crop lasted nearly 3 weeks without bolting. The large, deeply-veined leaves were a beautiful green.

This was 1959. Since then we have enjoyed delicious home-grown fresh spinach every season. In 1960 we gathered our first meal of these tasty greens on June 16. In 1961 it was June 28, and in 1962 we were cutting our first leaves on June 26.

Our spinach has been so prolific that our friends now enjoy it with us. One of its virtues is that even though planted late, it is earlier than any other vegetable, thereby making it a special treat when we are hungry for something home-grown.

Here is the summing up of our success:

1. The small area could be more readily dried out and tilled in a much quicker time than the large garden, thus giving us a head start on the spinach.

2. The shade from the tree kept the ground cool during the hottest part of the day. This preserved some of the moisture in the soil, while allowing enough sun for good growth.

3. We were close enough to the house to water the plants whenever necessary.

4. The rotted mulch enriched the soil organically. It helped fertilize the plants and changed the soil structure from resistant clay into rich black earth that dried more quickly, yet did not bake hard.

5. We planted Dark Green Bloomsdale, a variety resistant to bolting in hot weather, therefore more adapted to our climate.

6. We removed the mulch instead of planting in it in order to warm the soil for as early planting as possible. We then put mulch on again after harvesting our crop. This enriched the ground for the following year.

In gardening problems there is usually a simple organic answer, quite often found by persistence. Occasionally, however you will find the answer, as we did, by chance.

—Inez Grant

SQUASH

Summer Squash . . .
(and summer not)

All squashes are tender, and really should not be started until the soil warms up thoroughly. Sometimes I risk a hill in early May, which may often be cold and wet here (near Boston). I lay mulch lightly over the hill, or else use a glass jug cloche-style with the bottom knocked out. Still, germination is usually poor, growth slow, and the first fruit picked only a few days before that from the main planting, which is made about May 30.

Even before my first group of hills has matured, I plant again. The second planting will come along faster in the heat of midsummer, and when they in turn mature, I pull up group one. Squash plants are capable of producing all season until frost, but older plants, weakened by heavy production, are more vulnerable to disease and attack by insect pests. They're unsightly, too. And it can't be said too often that the first fruits are always the finest. Ideally, summer squash plants or vines should be pulled up after 4 weeks of bearing. (Cucumbers, too.)

To stimulate rapid, luxuriant, leafy growth and heavy bearing, squash needs lots of rich nourishment. Rotted manure can be dug into the hill before planting, with more manure or compost spread later about the young plant as a mulch. In a cramped garden, like mine, where hills are 3 feet apart instead of the usual 4, extra-rich feeding should be given; for example, I apply some liquid manure just before each plant starts to flower. Squash needs plenty of moisture, but mulched squashes rarely need to be watered.

Even bush-type squash takes up a sizable space. This can be utilized for a second crop—radishes, for instance, which also help to repel squash bugs. In midsummer, seedlings of other vegetables for fall use can be transplanted within the shadow of a mature squash plant that is soon to be pulled up. The seedlings are sheltered from the sun until they get established, then later take over the vacated space. Lettuce is especially appropriate for this intercropping, a dozen seedlings (some to be used half-grown) can be accommodated in the space, compared to 4 or 5 of broccoli or cabbage.

Pests: The main insect troublemakers are the same that attack related crops such as cucumbers and melons. (1) The striped cucumber beetle which spreads bacterial wilt may appear as soon as the sprouted seedlings. They're a good reason for sowing each hill thickly, then when plants are 4 to 5 inches high, thinning to one sturdy specimen. Wood ashes or ground limestone are safe repellents. It would be surprising if these beetles didn't show up at some time or other; just as surprising if healthy plants in healthy soil succumbed to their attacks.

(2) The squash bug lays reddish-brown eggs on the inside of the leaves. These clusters should be picked off and crushed before the grey nymphs

Feed summer squash liquid manure just before flowering to stimulate luxuriant bearing.

hatch out. The brown adults can be picked by hand or trapped under a board. Plant the above-mentioned radishes nearby, or strong-smelling nasturtiums.

(3) The squash borer—a white, fleshy grub—works inside the stem near the base. The stem can be slit and the borer taken out and killed. To encourage formation of new roots, mound dirt over leaf joints.

—Ruth Tirrell

Remedies for Squash Pests

Squash, melons, cucumbers and related crops are victims of the same trio of main troublemakers — the squash bug, the striped cucumber beetle, and the squash borer.

The squash or stink bug is especially fond of summer squash, laying reddish-brown eggs on the insides of the leaves. Pick off and crush these clusters before the gray nymphs hatch out. Older brown adults can also be hand-picked or trapped under boards set beside plants and turned over each morning to clear ambushed bugs. To deter from plants, grow some radishes or strong-smelling nasturtiums nearby.

Striped cucumber beetles, which spread bacterial wilt besides their other damage-doing, appear as early as the sprouted vegetable seedlings. They are a good reason for sowing each hill thickly, then thinning to one sturdy specimen when plants are 4 to 5 inches high. A ring of radishes around each plant also helps repel the beetle, as does planting a few nasturtium seeds at the same time you sow a hill of squash. Another deterrent is a sprinkling of wood ashes or ground limestone about the base of the plant and on the foliage. Put it on when leaves are wet and repeat after rain.

The squash vine borer works inside the stem, often causes the main root to die. To combat this culprit, encourage formation of new roots by mounding earth over several leaf joints. Start vine crops as soon as possible after soil warms up. Well-established plants are better able to resist both the borer and squash bug, which appear a little later than the striped beetle. In infested vines, slit the stem with a sharp knife where a small pile of "sawdust" shows the borer is present, and dig it out. Cheesecloth tents or plastic netting covers may also help curb all three insects.

By the Yardful

At my southeastern New Hampshire garden I start plants from seed about May 30 by working up a series of hills, stirring a generous amount of well-rotted cow manure into them. You might say the hills are mostly cow manure. Squash like a rich, light soil—preferably on the sandy side so it will warm up quickly—and a pH between 6.0 and 6.5. The bed should be well-drained but supplied with moisture and enough humus to hold it. Dig two or three spadefuls of rotted manure or compost into each hill. In heavy clay soils, add the same amount of sharp sand.

Space hills approximately 4 feet apart, and plant about 6 seeds at the top of each one, patting them in firmly. After seedlings have emerged and are established, I thin them down to 4 plants per hill. From then on, about all I do is give them a good soaking every two weeks with water pumped from a nearby stream.

When the plants are about a foot high I mulch them with hay accessible from an adjacent field—and fortunately cut just about the time mulching is needed. I spread it about 4 feet around the plants so that they will continue to benefit even when at full growth. Water is essential to both the growth and fruiting of plants, and the mulch of course helps to conserve it. Despite the drought experienced in our area last summer, my plants produced lavishly.

It is always a great surprise to me to see how quickly these squash grow. One day I'll see stubby "fingers" of fruit which grow into harvesting size almost overnight. A secret of tastiness is to pick them before they get too large—10 to 12 inches long is ideal. Picking will also encourage more to come along. Letting them grow too long also results in tough skins.

About mid-July I apply a second feeding of cow manure and stir it into the base of the plants, first gently removing the mulch. Year after year I'm also amazed to find these plants almost completely insect-free. This I largely attribute to organic gardening practice. Occasionally, I inspect the plants and find some wilted leaves, which I remove immediately.

One hill of 4 plants is about enough to keep a family of 4 generously supplied with squash until frost. Four hills, or 16 plants, yield an ample harvest for canning or for supplying the neighbors. To keep up the family's culinary interest, in addition to the usual boiling and baking of squash, you can fry them. Cut in slices ¼ inch thick, dip into egg batter and cracker crumbs, and brown both sides. Ready to grow—and serve—some this summer?

—Walter Masson

One Squash Makes a Truckload!

Talk about big—nobody comes close to some of the eye-popping, scale-busting vegetables organic gardeners have been growing lately.

Can you imagine calling a tow truck to haul a *single* squash out of your garden, for instance? That's what Bob Fox up at Ravenna, Ohio, did when his record-setting beauty weighed in at a hefty 250 pounds 4 ounces.

Actually, there's good down-to-earth methods — not inflation — involved in the results Fox has been getting. (The vine his 300-pound-plus specimen grew on also produced two others weighing 265 and 249 pounds—a total of over 815 pounds from one plant.) To grow squash and pumpkins in the large-economy-size class, he first advises testing the soil to see where pH and fertility levels are. Start plants inside about April 15th and set out May 15th. Keep a basket beside each hill for protective covering when late overnight frosts are expected. Make hills of old manure—he uses aged sheep manure—containing no straw, two feet in diameter and one foot deep. Mulch with straw around the base of each hilled plant about the middle of June.

Small squash should set by the second week of July, says Fox, who keeps a hive of bees within 10 feet of his patch, and urges other gardeners to do likewise. When the fruit reaches the size of a basketball, water 3 to 4 times a week unless enough rainfall

has done it for you. And to build his soil for such weighty production, Fox covers it every fall with 3 inches of manure, disks that under and then applies a 2- or 3-inch layer as ground cover until spring. He adds bone meal and greensand or potash rock when turning this under ahead of hill planting those Hungarian Gray squash. For insect protection, he dusts with wood ashes, plants garlic around the base of each plant, and tries a little hand-picking.

Adds prize-winner Bob Fox, who donates many of his big squash to old-age homes where they're converted into hundreds of delectable pies, "I've been busy with school children and 4-H'ers, showing them the pumpkin house I've built in the backyard and teaching them how to grow my favorite vegetable right!"

—M. C. Goldman

Mulching Makes "A Squash Queen"

Spreading fresh hay one foot thick around and between my vegetable rows has earned me the title of "Squash Queen" here in Traverse City, Michigan.

One year I harvested 633 pounds of squash from three seeds—one squash weighing 71½ pounds, while this past season, 8 vines produced over 1,200 pounds.

The big advantage in using hay before it is thoroughly dry is that it can be molded around plants easily, and another is that if it's alfalfa, the leaves stay on intact while dry hay sheds its leaves when handled.

In planting, I simply pull the mulch aside and drop the seed without any special soil preparation—and I've yet to see a bug. Earthworms had the fall and winter to make the soil friable—

protected by the deep layer of mulch. A neighbor who sprays poison on his squash vines to kill bugs did not have squash to harvest after all his efforts due to the bugs and dry weather.

Mulching not only takes care of the bugs but the moisture also, keeping topsoil from drying out. Occasionally when I make a soil test, pH is where it should be for garden fertility. Bone meal and cottonseed meal broadcast at the rate of 5 pounds to each 100 square feet should offset any nitrogen demands of the decomposing mulch materials—including cut-up corn stalks and vegetable vines. I'm sure a shredder would be nice but I get good results without one and have less work, I believe.

Besides the Blue Hubbards, I grow several other varieties and at the present time like best the sweet potato squash Eat-All that has hull-less seeds of high nutritional value—35 percent protein, 40 percent vegetable oil. I peel the outside skin and slice it like a cucumber directly into a pan on top of my range, in a little cooking oil. The squash browns nicely, is done in 5 minutes, and is simply delicious. I sell most of my large squash around Easter time—squash goes so well with ham—and for a premium price. Farmer Seed & Nursery Company of Fairbault, Minnesota, has the Eat-All variety.

I have also found that baling damp hay is good practice, as the hay will heat and spoil, killing any weed seeds, and that later layers of any thickness desired can be peeled off when needed to discourage any weeds that may have the idea of coming through the mulch. I also use some sawdust for mulch, as well as fall leaves which are all raked and placed on the garden before the snow flies.

I like the method of growing potatoes on top of the mulch that had all winter to decompose, and covering them with hay. In a test of mulched versus unmulched potatoes, I found that the unmulched had bugs and the plants did not have the dark-green foliage that the mulched ones had.

—Helen F. Miller

SUNFLOWER

Something for the Birds

One way sunflowers benefit growers—gardeners as well as farmers — is by providing a preferred food supply for birds, who in turn do a great job of assisting with insect control. If you want the birds around to lend a helping bill when your vegetables, fruits and flowers are maturing, let them know they'll find food at your place all year round—before and after the bugs. And one of the surest ways is to grow some giant sunflowers for the chirping pest fighters to harvest for themselves.

That was the idea D. C. Marshall of Manhattan, Kansas, had in mind. "The first year I tried growing sunflowers," he writes, "I planted a few and harvested what I thought was quite a lot of seed. When winter came and I began feeding it, the birds ate it so fast that the supply was gone long before the winter was. The next time I said to myself, 'I'll show them,' so last year I planted a row some 20 rods long. The crop was wonderful. The plants came up and began reaching for the sun. Up and up they stretched and the stalks swelled larger and larger to support the huge leaves and the enormous bloom that burst forth in midsummer.

"As soon as the seed formed, the race was on. The birds thought they should eat it right away. I thought they should save it for winter. We compromised. I harvested the largest heads and left the smaller ones for them. The result was I had a bushel and a half of seed, they had about 50 heads to take care of, and a grand time was had by all.

"The next problem was what to do with the stalks, which were as large as small trees. They ranged from 12 to 16 feet in height and two or three inches in diameter. They would make a lot of humus if they could be returned to the soil, I decided. The best way seemed to be to plow a furrow, then cut down stalks and lay them in the furrow and cover them with the next furrow slice. This was repeated over and over until all were buried. The stubs and the roots were left; they were too big to plow under with a small plow so we pulled them, stacked them into a compost pile, and let nature take its course.

"Checking up a few weeks after plowing under the stalks, I found they were well on the way to being completely decayed. Well, I gained a lot of vegetable matter in the soil, I lost a lot of bugs and worms, and I gained a lot of bird friends who are staying around. Also, I am sure that I have the approval of God who places a value on even the lowly sparrow."
—Earle W. Gage

Shade for Young Fruit Trees

Two years ago I purchased a small piece of land which had lain idle for 50 years, since it was a part of a brickyard. The first summer I rotary-tilled it and scattered sunflower seed to provide cover as well as food for the

158

How wide can a sunflower grow—and high? Charles Cohick of Salladasburg, Pa., grew this one 27 inches wide and 20 feet high.

birds and to supplement the ration I feed a few chickens I keep. In the fall I turned it under, having left many of the sunflower heads on the plants.

Although this heavy clay is slow to respond, I set out a few dwarf apple and cherry trees last spring. As the season progressed, I found many of the trees surrounded by sunflower plants. Being a bird-lover, I decided to let them grow and take a chance on the effects the plants might have on the young trees.

Imagine my delight to find that the shade and windbreak they had provided more than overcame the moisture they used. The trees which had been sheltered showed twice the growth and health of the transplants that had gone it alone.

I had another lesson from all this in that the sunflowers had sprung from seeds buried in the ground all winter. These seeds had sprouted and were above the ground before the land was really workable. Generally the 24th of

May is taken as the earliest date to plant seed in this area.

Rest assured that when I set out any more young trees—be it spring or fall—I will surround them with sunflower seeds!

—H. Raymond Goode

Four Tips for Towering Sunflowers

If you've ever wondered how some gardeners have managed to grow those towering sunflower plants with phenomenally big heads, we've got a few of their secrets to share with you. Over the past years, the Organic Gardening and Farming Annual Sunflower Contest has made awards to more than a score of participants.

The folks who grow the real giants among sunflowers, the 20-inch-and-over beauties that have won contest places, offer these suggestions:

1.) *Variety choice is important.* If you're intent on producing maximum head size, it's essential that you plant one of the tall, single-heading varieties. Dwarfs, semi-dwarfs and the cheerful multi-flowering types are nice—but not for this purpose. Make your selection from the Mammoth Russian, the largest and most commonly grown variety, maturing in 80 days with husky stalks and big, thin-shelled seeds rich in both flavor and food value; the Manchurian, another tall, large seeded strain taking 83 days to mature; or the Gray Stripe, a later variety (91 days, with similar qualities to Manchurian and lighter, handsomely striped seeds.

2.) *Plant carefully and with an eye on the calendar.* Although later plantings may be successful, they won't generally give you enough growing time to mature the biggest potential seed heads. Working with your own area's frost dates, allow the full length to maturity or slightly better for the variety you plant. Sow seed ½ inch deep about the end of April or the first week in May.

3.) *Space plants well; cultivate or mulch to control weeds.* Keep your plants a good 6 to 12 inches apart in rows 24 to 40 inches apart. Crowding will impede, not further, growth. Weed regularly or mulch to prevent soil-nutrient robbing by competing weeds. Water as needed, especially during hot, dry streaks.

4.) *Fertilize generously and prepare soil in advance.* Test for pH; best reading is between 6.0 and 7.5. Loosen and enrich soil; if possible turn under a green-manure crop in the fall. Rotate planting sites. Sunflowers are heavy feeders, do well on manure (well-aged poultry manure especially) and liberal applications of compost. Rock minerals help, too. Use ground phosphate and potash rock greensand or granite. Many contest winners report their plants were grown (or came up as "volunteers") near or on compost heaps, or where these were previously located. Fred Lemoine of Maynard, Mass., 1962's all-time champion with a 24⅝-inch sunflower, says his winner was grown with compost made up of leaves, bone meal, rock phosphate and ground limestone.

SWEET POTATO

As Serious as a Bricklayer

When father begins his sweet potato bed in early spring, he takes his work as seriously as a bricklayer. He well knows that things must be done exactly right if one expects a good crop in fall.

According to him, the type of soil, rather than its fertility, is the key to successful growing. The soil in which he grows his sweet potatoes is a light sandy loam—just right for this vegetable. He has grown them on heavier soil about the farm, but did not have much success unless he kept the plot under constant cultivation. He definitely shies away from clayey soils.

Experience has taught him that they produce poor yields and misshaped potatoes. (He likes his large and straight.)

Father works his sweet potato plot as early in spring as possible. If tests show that the soil's pH is below 5.2, he adds just enough lime to bring it up to about 6.5, which is about right for this vegetable.

After spading deeply, he rakes the plot smooth, then marks off the rows that are to be 18 inches wide and 3½ feet apart. He then runs a small garden plow down the very center of each row, forming a 6-inch-deep trench. This he fills completely with a fertilizer mixture which consists of equal parts of well-rotted cow manure and compost, as well as liberal amounts of phosphate and potash rock. (He stays clear of fertilizers high in nitrogen, as they produce all foliage and no potatoes.)

Without mixing this organic material with the soil, he forms a 10-inch-high ridge, 18 inches wide, down each row right on top of the fertilizer. This is one of his most important steps in his culture of sweet potatoes. It prevents the roots from going too deeply. They remain in and around the fertilizer, and produce large, straight potatoes within the loose soil of the ridge. This fertilization method enables any gardener to grow a decent crop.

After the ridges have been properly built, they are left to settle until planting time in early May. Meanwhile, to save space in her garden for vegetables with a longer growing period, my mother plants her early-maturing vegetables, such as peas and lettuce, in the aisles between the rows. These will mature before the sweet potato vines begin to creep.

To prevent loss of plants due to frost, father does not set out his plants until at least 10 days after the last expected frost. This is around the middle of May in the Middle West, and the last of May in the East.

When planting, father uses a hoe handle to make 6-inch-deep holes every 18 inches down each row. He finds this deep planting necessary. Not only does it place the roots just above the fertilizer, but it gets them down deep enough so that the potatoes are sufficiently covered with soil as they grow to maturity. Each plant is firmed well, then given a good watering.

As soon as the vines begin to creep, my dad removes what vegetables are left in the aisles, then mulches the entire plot with plenty of old hay or straw. The plot then gets a good soaking. From here on in, he pays no more attention to the patch other than occasionally lifting the longer vines to prevent them from rooting at the joints.

—Betty Brinhart

Year-Round the Easy Way

I am not an expert but I have grown sweet potatoes for many years and love it, but I do it the easy way. I just threw up a small ridge, set my plants about a foot apart, and when they begin to grow I pull more dirt up making the ridge larger. If the vines start out to bother other things, the tiller takes care of them or I chop them off with a hoe.

When frost comes I pay no attention to it. A week or two later, when the ground has taken all the frost out of them, I dig them and throw them into baskets and set them in the basement near the furnace, then eat sweet potatoes all winter and the next summer till the new crop comes. I used to dig them early, try all schemes to keep them, but throw most of them out rotten before spring. However I learned you don't have to baby sweet potatoes if you don't dig them too early. Now, at the last of March, my sweet potatoes are just like they were when I dug them, not a single one has spoiled, and I expect to eat them all summer.

—D. V. Davis

Without a Green Thumb

"I don't have a green thumb," says Max Gerner of Miami, Florida, "just a little common sense. That's why I stick to strictly organic methods. I get better results with less work, and the food I grow is fit to eat —not full of poison."

161

Compost and mulch are the secrets of Max's success. Each year he tips a local tree surgeon a couple of dollars to dump a load of wood chips near the vegetable patch. These are given time to rot, then worked into the soil or used as mulch.

A compost pit receives table-trimmings from fruit and vegetables, and clippings from lawn and trees. No animal matter is added to the compost. Max occasionally spreads a little chicken manure around the fruit trees.

Anyone who has ever grown white potatoes should remember one important difference when planting sweet potatoes: You plant only the sprouts of the sweet potato, not the cut-up tuber as with white potatoes.

Sweet potato sprouts may be obtained in the following ways:

1—Buy them from a nursery.

2—Set a sweet potato in a dish of water and let it sprout indoors.

3—In early spring, plant a whole potato (or a large one cut in half the long way) in a cold frame. Cover the tuber with about two inches of sand or light soil. Keep the sash of the cold frame raised enough for ventilation except when freeze threatens. Water the bed very lightly. Sprouts will begin to appear in 10 to 12 days.

4—Break off tips of growing vines and plant them. (This is Max Gerner's method.)

While the sprouts are developing, prepare the bed for them.

Max uses plenty of compost and wood chips in his sweet potato patch, but no manure. If sweet potatoes receive much nitrogen they run to vines instead of making nice, fat potatoes.

He plants the sprouts in ridges about 12 inches high and 3 feet apart.

Sprouts should be about 10 inches long when they are planted, and set about 15 inches apart in the ridge. Place the sprout deeply enough in the soil so that some leaves are covered, and firm the soil around the plant.

For the first few weeks, keep the bed weeded. Before long the vines will spread so thickly they'll smother the weeds. They'll then require only one other very easy bit of cultivation: The vines should be lifted and loosened every few weeks.

—Jeanne Wellenkamp

Starting in Sphagnum Moss

Since sweet potatoes require a minimum of 120 days to mature properly, it is very important that the plants start growing as soon as planted in garden. Tubers grown in a hotbed or greenhouse under two inches of light soil, kept moist, will produce rooted sprouts—often termed cuttings or slips. These sprouts can be bought from seedhouses, although I prefer to grow my own. I have found that the best way to get them started well is to pull the sprouts as soon as they have two leaves, wrap the lower part in sphagnum moss, and then set them back in the hotbed. The roots grow out from the main stem and become embedded in the moss. This should be done at least 3 weeks before time to set in garden. When the weather climbs above 50 degrees, I take the cover off the hotbed, and expose the plants. This hardens them and the shock is not noticeable when plants are set.

Use a rounded stick, like a broom handle, to push the roots of the plants 4 to 5 inches deep. Water after planting to settle the roots. Remember to keep the garden area around the plants free of weed growth until the vines shade out weeds themselves.

I have been growing sweet potatoes in Michigan for 8 years. I have always had fair yields, but last year, I hit the jackpot! This is how I did it: by knowing the secret of growing sweet potatoes was to get them started as rapidly as possible after setting them out in a garden I pulled off 24 sprouts and wrapped them in sphagnum moss, as described. I planted 3 rows of small-stem potatoes, and put the ones wrapped in sphagnum moss in the middle row. They had the same care, which was to keep the weeds and grass pulled out until the plants started to vine.

I could hardly believe my eyes when I harvested them. From the two rows planted the usual way, I got 54 pounds. From the middle row, I got 82 pounds of number-one grade sweet potatoes. This is almost 4 pounds per plant. By starting plants wrapped in sphagnum moss, you get about 3 weeks' jump on the season.

—B. B. Braden

Sweet Potatoes in January

In January, when gardening is at a low ebb, a sprouting sweet potato

B. B. Braden writes that "you get a 3-week jump on the season" by starting sweet potatoes in sphagnum moss. Yield was also 50 percent greater for sphagnum-started crop—almost 4 pounds for each plant.

brings a touch of new green to the porch windowsill, and becomes the very first token of the following summer's garden. We've found all that we need to get started is a potato of the flavor and texture that pleases us, 3 or 4 toothpicks, and an attractive container, almost full of water.

After choosing a perfectly sound sweet potato from the previous year's crop, I place it root end down in a deep container about two or three inches larger in diameter than the tuber. In order to keep about two-thirds of the potato up out of the water, I insert toothpicks horizontally into it in several places. The ends of the toothpicks rest on the edge of the jar, supporting the potato in the desired position. After several days, strong, white fibrous roots begin to show in the water. It takes a little longer for the tiny dark-green leaves to begin to unfold at the top of the potato.

Sometimes I let the vines grow as long as 20 to 24 inches. This gives a fine decorative effect, but it seems to me that sturdier sets are grown if the sprouts are removed when 3 to 5 leaves have formed. These sprouts, sometimes showing a few fine roots of their own, can easily be broken off

the sweet potato.

Placed in another container of water, they'll grow more roots and possibly a few more leaves. The long vines may be cut into lengths of 8 to 10 inches and put in water to grow roots, too, but this takes a few weeks. Late in April or early in May, I set the well-rooted sprouts in a trench in the cold frame where they harden until all chance of frost has passed. It's simple to protect them in the cold frame from a too-hot sun during the day or too-low temperature at night.

At about this same time, we prepare a couple of mounded rows of fine, rich soil in which to plant our sweet potato sprouts—or sets—when the weather is warm. Sets may be planted about 12 to 18 inches apart in the row when it is safe to plant lima beans or tomatoes. Heavy-feeders, they grow well when they have sufficient moisture. Mulching with compost helps fill both of these requirements.

To start sprouts for the quantity of sweet potatoes we need for our own use, as well as for gifts, just two medium-large, sound tubers are plenty. Each provides us with 25 to 40 sprouts. And each sprout, in turn, will grow into a plant that will produce several pounds of sweet potatoes.

SWISS CHARD

Drought-proof and Frost-free

All the greens are superb sources of vitamins and minerals—but as far as I'm concerned none have as many advantages as Swiss chard. Spinach, for instance, is a cool-weather vegetable, for it goes to seed in hot summer temperatures. Kale frequently becomes tough. Parsley has such a pungent flavor that you can use it only in small amounts. Beet greens are unusual in that you can use both the tops and the roots, too. But once they are pulled up, that's the end of your greens.

Swiss chard, on the other hand, defies both frosts and hot weather. You can plant it as early in the spring as you can get the ground ready, and it will produce until late into the fall. In some sections of the country where the winters are mild it will even produce all winter long. Here in northern Ohio, the tops die down in the wintertime but with a little protection— straw, hay, leaves or even snow will often do the job—the roots will live on and early in the spring start producing greens long before the earliest lettuce is ready. I gather these tender, succulent leaves and use them in salads. (One word of caution: It's best to add the dressing just before serving or leaves have a tendency to become dark and wilted.) Swiss chard's extremely high vitamin A content and considerable amount of vitamin C make it an impressive addition to the early-spring diet.

Crops like corn, potatoes, melons and squash require a lot of ground for their yields, but my 10 feet square of Swiss chard produces more than my family can use. This small space yields more vitamins and minerals than many another section of a hundred square feet.

Each summer as I compare notes with other gardeners I listen to tales of troubles with drought, or too much rain, corn with either no ears or ears that refuse to fill out, tomato blight, cabbage worms, bean beetles, and peas that mysteriously die. After fighting the various problems of different plants, it's a relief to find one that seemingly has none.

Swiss chard is as trouble-free a vegetable as you'll find. I simply toss a few seeds in an inch-deep furrow early in the spring, mulch it generously, and forget it until I want to use it. It does well in rich soil, but I've had good results in poor soil, too. It does well if you have a rainy summer, but last year's dry summer didn't bother it a bit. If Swiss chard could be said to need any special treatment, it would be the addition of lime to very acid soil.

I have never known a bug or worm to attack my chard. If small brown spots are noticed on the leaves, it's far more likely to be a soil deficiency than an insect infestation. Spots along the edge of the leaves usually denote a potash deficiency; in the middle of the leaves, a magnesium deficiency.

Dolomite limestone will correct the magnesium problem; wood ashes, granite dust, or greensand will supply the potash.

Swiss chard is a member of the beet family. Instead of the bulbous roots of the beet, however, it has a tremendous root system that thrusts its tentacles deep into the soil. Such is the penetrating power of the roots that Swiss chard would make an excellent soil conditioner, breaking up hard pans, aerating the soil, adding quantities of humus from the root system alone. Try it if you have a piece of ground that needs rejuvenating.

Finally, getting children to eat vegetables like Swiss chard calls for a little ingenuity—but who can resist a hearty, savory vegetable soup? Chard not only adds tremendously to the vitamin value of the soup but enhances the flavor as well.

Next time you plant a garden, save a small corner for Swiss chard. You'll be glad you did.

—Lucille Shade

The Versatile Swiss Chard

If you have a small family or a significant lack of space, try growing some chard as a foliage background in your flower garden. One variety of this beautful plant is the rhubarb chard, an eye-catcher with its reddish leaves and brilliant crimson stem. I prefer the regular chard, Lucullus or Fordhook Giant, for general eating purposes. These have dark-green, glossy leaves and white stems.

Start cutting your leaves as soon as they are long enough to make it worth your while. Boil the leafy part just as you would spinach and cook the fleshy white stems and serve like asparagus. With chard you get the bonus of two vegetables in one and the added bonus of no metallic "aftertaste" so frequent in spinach.

This is a cut-and-come-again vegetable, so after a few favorable days a new growth of delicious leaves starts appearing. And this cycle of cutting and regrowth continues all summer and into the winter. In my New York state garden we harvest our chard almost until Christmas. It survives severe frost and ice storms with remarkable fortitude. A blanketing of straw makes it more accessible after snow starts falling.

Another good feature is its ability to thrive during scorching summer days. Unlike some other types of greens, it doesn't turn sulky and wilt under sizzling suns, provided it gets sufficient moisture. I grow ours in a mulched garden thereby eliminating the moisture problem entirely. In an extreme heat wave, leaf growth may be slowed and the leaves smaller; if this happens, new seed may be planted to be ready for harvest after a month to 6 weeks.

If you leave your chard plants over the winter they will provide a cutting or two before bolting to seed. This gives you a head start on the season while waiting for newly planted chard to establish itself.

Soil requirements are simple. Any rich, mellow garden soil will do. It does best in a sunny, open, well-drained location. Lack of magnesium or potash in the ground may result in brown spots on the leaves. The cure for this—bone meal, wood ashes, or ground potash rock added as to your need. But actually chard isn't a gross feeder and can get by on pretty slim rations with fair yields.

On the nutrition side of the ledger chard ranks high as a supplier of vitamin A and is an excellent source of calcium, phosphorous and iron.

TOMATO

I'm Growing 8-foot Tomatoes!

Talk about tomatoes! The blossom hands (flower clusters) on plants in my Florida garden grew 12 inches or longer, and each set 11 to 13 tomatoes. Then the ends of the blossom hands started to grow and put out leaves, and have been blooming all along the new stem. Some are 3 to 4 feet long, and one sprouting hand that I tied to a stake reached 6 feet in length by mid-February, with bunches of tomatoes along both sides. Even leaves have been putting out suckers from the top joints and blooming—and it looks as though I'll have fruit on them.

In order to grow successive crops on the same soil, I keep enriching it. In the fall I spread about two inches of compost over the plot and dig it into the soil. When I plant, I use a mixture I toss together in a wheelbarrow— some well-rotted cow manure, ground phosphate rock, granite dust, a commercial organic blend, dolomite, bone meal and tobacco stems. The results I've had with all my crops—not just tomatoes—are tops. If it means climbing ladders to harvest yields like this, I'm ready!

—Paul A. Mahan

Getting a Jump on Spring

Say hello to spring this year by starting your tomatoes in their own individual greenhouses. For years we picked our first tomatoes and bell peppers around July 4th. But now, thanks to our one-to-a-plant green-houses, we're picking them a month earlier. What's more, those sturdy, handy and useful greenhouses can be used repeatedly, year after year.

Ours are made from 18-inch chicken netting, cut with wire nippers and made into a circular enclosure, laced together with heavy twine. A piece of cardboard is lashed to one side to cut down on the too-intense light, and finally the unit is covered with heavy plastic, the kind we have left over from covering the windows in the winter.

Because we seldom have a killing frost in our part of Texas after Easter, we always put our tomatoes, peppers and eggplants out in the garden the Monday after Easter—that is, until we made a dozen individual green-houses. These greenhouses enable us to put out a few tomatoes and peppers several weeks earlier for early eating, while the rest of the plants are left in the hotbed until our area is frost-free, when they are put out at the regular time for canning and freezing. But those early few grace our table and make wonderful eating.

The early tomatoes and peppers are planted in the same way as the others, in soil rich in compost and watered in place with manure "tea," that seems to help the shock of transplanting. Next, the individual green-houses are placed around each plant, and the dirt is mounded up a little way around the bottom to cut out drafts, and to keep the greenhouse from blowing away in the occasional

Burgeoning young plants grow through the tops of their individual "greenhouses."

high winds we have in spring.

The greenhouses must have coverings at night. We use various things; grocery sacks, one inside the other, may be slipped down over each greenhouse, and if I am out of sacks, a basket or old dishpan covers the top.

It is only the work of a moment the next morning to take the coverings off the greenhouses when I go to feed the animals.

If a cold north wind blows, let it. The air inside these greenhouses is warm and humid, and the soil doesn't dry out nearly as fast. If it does get dry, I use more manure tea instead of water.

Even after frost is past, we leave the greenhouses around the plants for a few weeks to keep the soil warm. This helps the first fruit buds stick, and insures an early crop. The plants grow right out the top of their greenhouses, and when we're sure spring

is here to stay, we carefully remove the greenhouses and store them for another season's use. The wire, plastic and twine are all leftovers from previous projects, so the greenhouses are inexpensive as well as easy to make. If you're looking for early tomatoes this year, get started on your individual greenhouses.

—Margilee J. Rozell

Early—Despite the Season

Nobody said it couldn't be done —that is, to set tender plants in the home garden two or three weeks before recorded late-frost dates. The charts merely advocate waiting until after those danger dates have passed.

That's easy advice to follow on Long Island, for instance, where the long growing season assures ample time for the gardener to enjoy the fruits of his labors. But what about the so-called "ice-box" areas of many northern states—places where only a few degrees difference can mean unpredicted frosts—and frequently do?

For some years, I have kept close watch on weather habits here in one of New York State's colder regions, where readings in the thirties are not so unusual every summer.

Properly acclimated plants are not bothered by chilly nights and light frosts. It is those unhardened, tender little transplants taken from an east window or bought from a city shop, that one has to fear for.

I was convinced it was possible to pick ripe tomatoes in July. If all a plant needed was some rigid hardening to save it from a light frost, it seemed to me that a toughened one, with a cover, should withstand a short freeze or a hard frost. This would mean growing my own plants or finding a commercial grower who carefully exposed his flats to the elements as much as possible.

So I bought some sturdy, well-conditioned tomato seedlings, then set them deep in the soil on the second of two balmy days in mid-May. Each plant was covered with a bushel basket—the kind with openings between the slats so light could enter—pushed close to the soil and weighted with a brick or stone against sudden wind or stray animals. Still, when snow or very cold nights came, I was apprehensive over their survival. On May 24, our official low temperature was 19 degrees! But no damage occurred.

Whenever readings hit 40 degrees or higher, the baskets were removed for a short time. If winds prevailed, the leaves quickly wilted, but after a few days the exposure time was lengthened. With ample moisture and a good mulch, the rugged little plants came through great. Some were blossoming by the time covers were left off permanently.

That first trial year, the tomatoes produced heavily, but did not ripen earlier. I had chosen two late-maturing varieties—Rutgers and Big Boy. The following year, with the same basket procedure, early Fireballs were used. Small to medium-sized fruit ripened in mid-July on this compact plant that produces abundantly and requires minimum spacing.

Other years and other varieties followed, but none more satisfying than 1966 when Moreton Hybrids were seeded in mid-March, transplanted into 3-inch peat pots when two inches tall, and kept near a window in a very cool room. Meanwhile, the tomato area in the garden was prepared as usual, by working rich compost, wood ashes and bone meal into the soil for improved fertility.

This time, watching weather patterns closely, I set my plants out on May 18th, each under a hotcap with a basket over that for the first few days. Later, the baskets were removed and used only when a freeze was likely.

Soon the tomatoes were pushing hard. The tops of the hotcaps had to be removed, leaving only the lower portion as protection around the bottoms for awhile longer. Those over the vines were eventually split for ventilation, then left in place until the plants were crowding for space.

Not only did these tomatoes give July fruits, but the Moreton Hybrids—medium to large, sweet-flavored and very heavy bearers—provided a total yield of over 200 pounds from 8 plants. They were not staked, but this being such a vigorous variety, I spaced them two feet apart in rows 5 feet apart.

One important step in forcing earlier ripening of tomatoes is pruning. It's a trial-and-error process, depending on weather. Heavy spring rains cause lush growth, with little progress in fruit development. When there are enough blossoms and small fruits for anticipated needs (and the larger ones have not yet begun to color up), I clip every end back to one leaf above the second fruit spur on the stem. Then I remove most of the sucker shoots that tend to shade the first-formed fruits, leaving only an occasional cluster of buds to develop small green pickling tomatoes. In all cool regions, this pruning should be done early in July.

—Ellie Van Wicklen

Cold Treatment

To grow early tomatoes give the young plant the cold treatment recom-

169

mended by Michigan State University researchers.

The cold treatment consists in growing the plants for 3 weeks at night temperatures of about 50° F. to 55° F., beginning after the seed leaves have unfolded. This can be done before seedlings are picked off, during this operation and afterward.

Such cold treatment has been shown conclusively year after year to precondition the blossoming closer to the ground level on sturdier plants. The plants have stronger side shoots and are hardier against the hazards of transplanting and early growth than transplants which have been grown indoors continuously as usual at higher temperatures.

—Gordon Morrison

The First Tomatoes in Town

I planted Firesteel tomato seed in a cloth-covered cold frame last March. Plants produced in this way will stand light frosts and are quite hardy.

When the plants were 3 or 4 inches high, I transplanted them into deeply mulched beds; the leaf and straw mulch was from 10 to 12 inches deep. The plants were set in depressions which I made by spreading the mulch apart all the way down to the soil level at a distance of 3 feet apart each way. The roots were set in the soil and the mulch was then pulled back around the plant stems, leaving the leaves and bud out above the mulch level. This left each tomato plant in its individual depression in the mulch bed. Then, over each depression I placed a windowpane. Each plant was thus left in its individual hot house.

By the time the plants were pressing their leaves against the glass it was warm enough to remove the windowpanes. At this time I levelled the

mulch around the tomato plants, eliminating the depressions, or practically so.

The only other work I did was to start picking tomatoes early in June. The harvest was abundant all summer.

I used no special fertilization for these tomatoes. Our entire garden has been heavily treated with phosphate rock and granite dust, and our continuous mulch system of gardening seems to supply everything that plants need.

I am planning to start my plants in February this year. The hay depressions covered with windowpanes will produce plenty of warmth for an earlier start, I think. Anyway, we shall see . . .

—Vernon Ward

In a Jug

I raised my own plants and 6 weeks before the last frost date here I set them directly into my garden (April 18). Over them I put a gallon glass jug with the bottom knocked out, no cork in top. I planted them 6 inches below ground level and set the jugs down in the depression, after they grew to the top of jugs I pulled dirt in around the plants and set the jugs on top of the dirt again, this was to keep the starlings from cutting the young plants off. On May 30, the usual time to set plants in the garden here, I had blossoms on my plants, June 15 the blossoms were set. I used the same idea for my cukes and canteloupe and I found the first cuke set on June 17th.

The jugs came from hairdressers who seem to have plenty of them to throw away. I boiled the jug about 10 minutes until thoroughly heated through and plunged them into about 1 inch of ice water. They cracked and then I tapped the bottoms out with a wooden hammer. Some came out nice and smooth, (just the bottoms) others not so even. Now if someone would manufacture jugs with smooth edges to use this way I'd be all set. Jugs are much superior to hot caps because you allow air to enter through the top of jug and plants harden off easily.

—Mrs. Homer R. Sprague

Indian Method

Start in a part of the garden where no shadows from trees, buildings, or other objects will fall on the plants. There must be a full day's sunlight and warmth. Now dig a hole about 18 inches deep and as wide as a shovel can comfortably work in. Into the hole toss about 3 inches of corn cobs or chopped corn stocks.

Over this place about 2 inches of cow, chicken, or rotted horse manure. Dried manures may be bought at most feed or seed stores if otherwise not available. Next pour about 4 inches of soil from the compost pile or from the richest topsoil area of the garden over the manure. Warning! Use no commercial fertilizers in any manner to supplement this method. The Indians used fish, or deer or buffalo manure.

Now for the plants. These should be sturdy plants at least a foot tall. All of them should be given a good soaking in their pots or cold frames one day before transplanting into the garden. They transplant much better after having absorbed an ample supply of water. You may have misgivings at the next step but have courage. From each plant pinch off every leaf and branch except on the very top.

Place the plant on the rich soil covering the manure in the hole. With equally rich soil cover the entire stalk of the plant except for two or three inches of the tip. Do not be dismayed at the puny looking results of this moment. That tiny crown will soon burgeon into a vigorous vine.

Here is what happens over the growing period: The corn cobs and stalks suck up moisture like a blotter. The dampness reacts on the manure which produces heat. The heat from the manure warms the soil. The warmth of the soil encourages rapid and continued growth of the tomato plant. The roots grow downward toward the manure-impregnated soil and find a great store of nourishment. From then on the plants may almost be seen adding new growth.

Never set plants right on the manure. They will invariably die. Make sure there is a protective layer of soil between plant and manure.

Do not add anything on the soil surface of the plant except a mulch after the plant has made some growth.

—John Krill

Tomatoes Until Christmas

By using plastic . . . there is a way to save tomatoes from frost. Buy a few yards of wire-mesh plastic sheeting, the kind used for windows in poultry houses. Out of this, a few dozen thumbtacks, and some poles at least 45 inches long, you can make tents that will keep your tomato plants snug and safe from all but the black frost that kills everything.

Make your tents by cutting from the plastic sheeting 4 squares as large as the width of the material permits. Place them together so that they form one large square. Now, using the point where the 4 squares meet as the center, draw a 6-inch circle. Around the outside of the large square draw

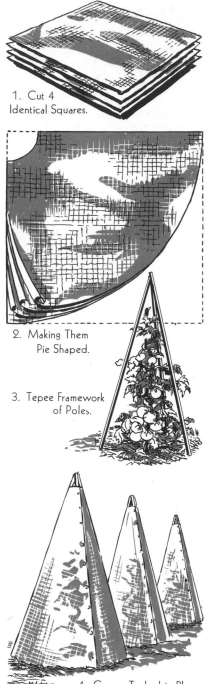

1. Cut 4 Identical Squares.

2. Making Them Pie Shaped.

3. Tepee Framework of Poles.

4. Covers Tacked in Place.

another circle. Cut out both of these, using metal shears or a knife. I simply laid my squares on the floor of the hardware store when I bought my material and let the clerk cut it, to the great interest of an increasing number of amused customers. You now have 4 pie-shaped segments, each of which will become a shelter for a tomato plant.

When frost threatens make your enclosure by driving 3 of the poles into the ground on 3 sides of the tomato, tilting them so that they meet at the center. Tie them together. Now wrap your pie segment around them and thumbtack both long edges together up one of the poles. Bank up earth or mulch at the base so that sharp fall winds will not be able to whip it away or frost creep in under it.

Light and air are permitted to enter this enclosure, but cold is kept out. I found it most effective to put on this winter "overcoat" when frost first threatened and to keep it there until really frigid weather ended all hope of harvest. When that time finally came I removed and stacked my tents in the tool shed. In the spring they again became weather lengtheners, permitting me to set out my tomatoes under them a good month early.

Last fall this device kept the Colorado frost from my tomatoes until the temperature dropped below 22 degrees.

—Dorothy Schroeder

. . . Or Cornstalks!

Borrowing a trick from our old-time Arkansas farmers, I use my cornstalks to give me tomatoes until after Christmas. Here's how I do it.

In mid-September I select two or three branches of my best Manalucie vines to layer them. I use the Mana-lucie for its fine quality—thick-meated, firm and flavorful—producing a strong, husky, sprawling vine that is noted for its resistance to disease.

I layer by cutting part-way through the stem and then covering the wound with moist, rich compost. Tomatoes often layer naturally, and it is a simple matter to produce strong plants in this way from growths that are already laden with blossoms and small fruit.

Later in the month, when the corn-stalks are shocked, I cut them up into 6-inch sections, mix them with cotton-seed meal and fill my hotbed with them. Next I give them a heavy soaking and pack them into a firm mass by tramping them down. They are moistened—but not soaked—every other day until they start to heat up, then covered with a one-inch layer of grass clippings or straw which is followed by 4 inches of my best garden soil.

Killing frosts come with mid-October in our part of Arkansas and, after listening to the weather reports carefully, I cover my tomatoes on cold nights, delaying as long as I can their actual transplanting to the hotbed. I can keep them going safely until mid-

Cross-section shows chopped stalks mixed with cottonseed meal which are topped by a one-inch layer of grass, plus 4 inches of author Croley's "best garden soil."

November here in the Ozarks under a heavy tarpaulin, but, after Thanksgiving, it is time to get the plants into the hotbed of cornstalks where a month to 6 weeks of gentle bottom heat keeps them growing and producing through Christmas.

—Victor A. Croley

Winter Tomatoes—Under Glass

Lots of glass facing south—that's about all you need to produce vine-ripened tomatoes over the winter, according to Charles F. Jenkins. Writing from his home in New Riegel, Ohio, Jenkins advises that "you are practically in the off-season, home-grown tomato business if you have a glass-enclosed porch or a solarium."

You'll have to maintain a 60-degree environment at night and 69 to 77 degrees during the day, using some sort of heating extensions. Jenkins plants in a mixture of ¼ sand, ¼ peat moss and ½ Heavy Ohio loam with a pH rating of 6.0 to 6.5.

The "right kind" of tomato seeds for indoor growing is important. Jenkins writes: "I found out the hard way that hothouse strains of tomatoes

Cutaway sketch shows tomatoes layered in compost after the stem was partially cut.

are the best in the long run. Hothouse tomatoes are bred to bloom and to continue growing as days get shorter. One year I tried the high-vitamin varieties which are wonderful outdoors, but under glass they are sad, to say the least.

Outdoor-type hybrid tomatoes grow in direct proportion to daylight, which means that in November and later, blossoms will be scant. One season I found an old standby, Rutgers Certified, will take off like a sprint runner in August and September, but when days shorten, it just stops growing."

In northern Ohio Jenkins finds that planting seed during the third week in June gives best results for the fall crop while planting the first week in December is best for the spring crop. He plants about 6 seeds to the pot, pinching down to the strongest seedlings after 5 weeks.

He supports the young climbers on baling twine suspended from overhead wires, gently wrapping the vine around the cord, and letting it up freely—and then back down again.

Indoor pollination must be done by hand because there is little or no air movement. Jenkins "lightly taps the flower stem when the bloom is wide open," which is a lot simpler than the electric vibrators used by the professionals.

—M. F. Franz

Tomatoes in December

Toward the last of October on the day the weatherman predicted our first hard freeze, we picked all the tomatoes, ripe, half-ripe and green, about 5 dozen in all. We solved the problem of what to do with them by spreading several layers of newspaper on a large table in the basement and placing the tomatoes on it, leaving space between each one. Once in a while I went down and culled out the soft or spoiled ones, but actually very few were lost that way. We ate the last ones during the first week in December.

—Jean Bible

The Hill System for Tomatoes

I was determined to build up my plot of gravel, clay and chalk so that it could grow tomatoes. To ready the plot I used a spading fork and dug holes about 18 inches across, mixing in one shovelful of rotted manure and a handful of ground phosphate rock. I mixed this with the soil like tossing a salad. My hills were 4 feet apart, but I had 3 rows, so I staggered them. The hills were prepared about a month before I planted to give the manure time to condition the soil.

Direct field planting is not generally feasible for growing early tomato plants because of late frost and cold soil. But for fall tomatoes it is best. Although you waste a little more seed and have to thin out extra seedlings, you practically eliminate any chance of damping-off and virus diseases which spread so readily in transplanting. Timing, of course, is very important when planting seed directly in the garden.

I chose the Marglobe variety which has been around since 1925 and has proven to be a tomato that can give you a good crop in the South even under adverse conditions. It is resistant to nailhead rust and resists fusarium wilt pretty well. I also used about half Rutgers seed because it was developed from a Marglobe cross. It ripens a few days later but has larger tomatoes and more of them.

After I planted the tomatoes, a few of the hills did not come up, so I did some transplanting on the first of July. My wife thought I had really "flipped" trying to get something to live and grow where it was so hot and dry; but I took the water spade (Hydra-Spade) and made a hole in the middle of the hill with water pressure, set the tomato in, and caved the sides in—and lost but 3 plants in the whole field. Then I used the spade to make a hole about 8 inches from each plant and about a foot deep through which to feed and water them. I had prepared two 55-gallon drums of manure water a month before, so started putting about a pint to the hill, then watered them again to dilute the manure water. I had mixed in some fish emulsion fertilizer, cottonseed meal and phosphate rock when I made the manure water. There was no other fertilizer used, and there was no mulch because I wanted to experiment without it to prove a method.

The tomatoes were not staked—it takes a lot of time to stake 75 hills, and the hot sun will scald them besides. I had no trouble with rot—it was too dry. Foliage was so thick there was no sunscald, and weeds were kept out with a hoe—no tilling, only hilling with the hoe. I had no trouble with pests of any kind, so there was no need to spray. The organic method proved again that a good healthy plant can pretty well take care of itself.

Support Your Tomatoes

You'll get better crops, more vitamins, easier picking and fewer insects when you grow your tomatoes on supports off the ground. The type of support doesn't seem to matter too much.

"When we set out to take the backache out of tomato picking, we went

Double hog-fencing is credited with 30 percent gain in yield, "earlier, better-flavored tomatoes, and easier picking."

on a search for hog netting," recalls Anoma Hoffmeister of Nebraska. "Or you can use any other cheap netting material available locally. Chicken wire or netting, however, is too light."

The Hoffmeister system calls for setting the tomato plants quite close together—18 inches apart in rows "so they hold each other up and do not slump down." The rows are completely enclosed in a double-fencing of wire netting strung on heavy posts securely emplaced.

The heavy-duty, double netting is strung up after the plants have taken hold and are thriving. The posts are spaced each 16 feet in pairs opposite each other, 18 inches apart. The netting should be rolled with the larger mesh at the bottom making it easier to pick the lower-growing tomatoes. And it should be securely fastened to the posts starting 6 inches above the ground and extending up for about 26 inches to make an enclosure about 32 inches high.

Taking the fencing down and stacking up the posts is an after-frost job, done without too much bother. The

Hoffmeisters stress that the "convenience during the summer far overshadows the work of erecting the fences and taking them down." They give the double-fencing credit for a 30 percent gain in yield and in "earlier, better-flavored, larger tomatoes and easier picking."

Trellis Technique

Arthur Langford depends on a lightweight but sturdy system of shoulder-high trellises to raise tomatoes on his Iowa farm. Large, heavy-bearing varieties are best suited to the trellis.

The rows should be set 36 inches apart and the plants spaced each 30 inches in the row. Where possible, run the rows north and south which will permit better light between the rows for uniform ripening.

Stakes should be put into the ground before the plants are large enough to require support. Drive them down deep so they will sustain a heavy load of tomatoes without tipping at a height of 4 to 5 feet. The stakes should taper to an inch diameter at the small end. Two-by-two

Shoulder-high trellises practically eliminate loss from sunburn and ground rot. The plants are spaced 30 inches apart.

stakes are also recommended for repeated use from one season to the next.

For horizontal supports, use small willows or sticks about the size of a cane fish pole. Some care is needed in nailing them to the verticals without splitting; slim box nails are recommended. The supports should be nailed to the stakes at even intervals, starting a foot from the ground and using 4 or 5 to the row.

As the plants grow, tie the main stalk to the stakes every 12 inches until they reach the top of the trellis. When they extend a few inches over the supports, trim off the end and keep any new growth cut. Mr. Langford stresses the vines heal quickly and grow new shoots.

Follow the Kniffin system of grape pruning, select vigorous side growths and tie them to the horizontal supports, suppressing all other competing side growth. Keep tying the side growths as they progress and trim them when they reach the top. Once a week, go over the patch and tie up any loose vines.

We have been using lightweight triangular racks made of 14-gauge galvanized wire to protect and train vegetables in our eastern Pennsylvania garden. Nine years ago, predatory vegetarians were making heavy inroads on our early spring plantings and we were forced to protect the tender young growths or lose the entire crop.

As the diagram shows, we made long, tent-like, triangular housings which were set over entire rows. The 14-gauge wire is easy to shape with an ordinary pair of pliers or nippers, and we have found that two people can make a wire frame or rack 30 feet long by 18 inches high and closed at both ends in half an hour. We are

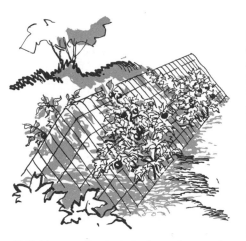

Light, 14-gauge galvanized wire frames keep predators away, and can be used to support and train tender young plants.

still using the frames we made 9 years ago but have extended their functions to include supporting and training young tomatoes.

We set a frame between two rows of young plants, taking normal care not to damage the tender growths. When necessary we tie the stems lightly to the frame. Otherwise, we merely guide and set the branches gently into place.

—Maurice Franz

Rack Method

I set plants in a row 18 inches apart. When they're growing well, I put up racks 8 feet long and 4 feet high, made of one-by-two-inch slats spaced a foot apart. Two of the upright slats are nailed a foot from the end and extend a foot below so they can be set in the ground. I lean the rack slightly and brace with stakes driven into the ground at an angle and nailed to the rack. I use pieces of old steel posts that have holes in them, making them easy to nail. Wooden stakes may also be used.

Racks and stakes must be set deep enough to keep them from blowing over. An inverted V-rack may be used and plants set on both sides, but single racks make hoeing and picking easier. When spring weather is quite warm I plant some seeds of Stone or Firesteel in the garden and reset when big enough. Those grow very fast and will bear late or until frost when the others are through. I have never had any trouble with disease, wilt or pests on my tomatoes.

As soon as side shoots appear on the plants, I prune them off, leaving the main stem and top which I tie to the rack. I usually prune off side shoots three or four times, training and tying main stems to the rack as growth progresses. I then let the plants go, tying as necessary.

—Charles W. Norris

Fenced-in Tomatoes

An almost continous crop of more and larger tomatoes without the pains of staking, tying or framing— how would you like that? What's more, our method keeps the tomatoes

Leaning racks produce tomatoes as late as Thanksgiving, with some lasting up until Christmas after they had ripened.

177

A fence 3 feet across requires 10 feet of wire; a two-foot-wide fence needs 7 feet.

off the ground without using straw, and gives us twice as many tomatoes from each plant.

It's done simply by fencing your plants within circles of ordinary 5-foot-high wire fence that you can get at your hardware store. The plants branch out inside the circle which supports the young tender branches, allowing them to get light and air all the way up and down and through the center. The tomatoes grow in more or less uniform size from top to bottom of the plant, and no cultivation is necessary once the fence has been placed. You weed by hand through it and also pick the tomatoes that way— which is easy on an aching back as the crop is distributed fairly evenly from top to bottom.

Some suckering will have to be done, but not as much as usual, and if you have bought the plants rather than raising them yourself, you may want to stick the suckers in the ground to try your luck with them. They won't be as large as the parent plants but often do well and bear quite a number of tomatoes themselves.

Now for your fences. For plants which grow unusually large you will need a circle about 3 feet in diameter, which requires 10 feet of fence. For medium-sized plants, a two-foot circle which takes about 7 feet of wire will be adequate. Be sure that the space between the squares is large enough to get your hand through, and leave enough extra wire to close the circle by looping the cut ends.

Now to prepare your soil. The ground should be cultivated a few weeks ahead of time, and some kind of organic fertilizer like rotted manure, compost, or bone meal spread over and worked in. Several years ago we were fortunate in getting a truck-load of rich river-bottom soil to add to our little garden plot, so we don't have to add much to it. By itself, our red clay would be a little too heavy and bake too easily for a really good garden.

The plants should be set in holes large enough to accommodate the roots comfortably and deep enough so that they can be covered to at least midway between the roots and the first branch. *Set the plants far enough*

The soil should be cultivated in advance, and enriched with compost or fertilizers.

apart for the light and air and you to get between the fences comfortably, about 4 to 6 feet. If this seems to take up a lot of space, remember you will only set out half as many plants as usual.

Before placing the plants, line the bottom of the hole with compost. Set them in, and water by hand before pulling the dirt up around them. This is more satisfactory than watering from the top after they are already covered. Your plants get off to a better start if you set them in the cool of the evening or on a cloudy day; but if they wilt a bit from the next day's sun —it won't hurt them.

For the first few days, water them any time they seem to droop. After they get a good start, watering need only be done occasionally during dry periods. We have found that the flat plastic hose with holes in one side that you lay on the ground along the rows allows a gentle seeping of water with no flooding and no evaporation. It is also quite light and easy to handle.

The fences are placed over the plants after 7 to 14 days' growth, and

Set the fences 4 to 6 feet apart for lots of light and air. Good spacing will also allow you to move easily between plants.

Anchor the circle securely with stakes if you are gardening in a windswept region.

before they get top-heavy. They are fairly rigid, but if your area has occasional high winds you may want to anchor them with small stakes driven into the ground.

—Jean Bible

New Tomato-Growing Ideas

New ideas in tomato culture keep coming up as fast as spring-sown seedlings. For example, R. F. Sullivan of Winston-Salem, N.C., tried growing tomatoes in wheat straw last summer. He bought two bales, laid them end to end with the twine still attached, and planted a half-dozen seedlings. "That straw was packed so tight that when I dug the hole for a plant it would close up tight again," he said. Still, it proved a better medium than the hard red clay in his yard. Sullivan added a little fertilizer and some pine needles on top to hold water. He lost one plant to a wind storm, but harvested dozens of tomatoes from the others, and was so pleased with the results that he may set out more bales for other vegetables this season. "Maybe this is a good idea for people who live where there's no soil, like in rocky places," commented Sullivan.

At Texas A & M University, another work-saving method has been developed—this one perhaps ideal for "weekend gardeners." Last summer, extension horticulturists Mack Fuqua and John E. Larsen set up tomato-growing beds which do away with watering and weeding. The bed's frame, they explain, can be made of lumber or other handy material to retain organic matter, and should be about 8 to 12 inches, about two feet wide and 20 feet long.

Mulches used in the frames are redwood sawdust, pine bark—any organic material that decomposes slowly can be used. These do not heat up after a rain and kill plants. Plastic film is laid under the shavings on top of the bed. In transplanting the tomatoes, a hole is dug to the plastic, and a small hole is punched in it. The cavity in the shavings is refilled with rich soil and then covered with mulch.

The tomatoes are called "carefree" because once transplanted into this rich soil, they do not need watering or weeding. Soil moisture vapor condenses on the plastic and keeps the soil underneath moist. The bed is kept damp by capillary movement of moisture from under the plastic to that in the bed. Plant roots grow in the rich soil mixture in the bed, under the plastic, and in the sawdust or other organic mulch.

"Rather than punch a hole in the plastic, another way is to cut a two- to three-inch hole in the plastic," adds Dr. Larsen, who has continued testing the idea developed by the late Mrs. Mary Robinson of Troy, Texas.

Dr. Larsen also cites a good way to discourage damping-off of young seedlings when planting tomato seed: Fill the container with soil mix to within ¾ inch of top. Place seed on top and then apply about ¼ inch of sand over it. Always water plants in the morning so that both plant and soil surface will be dry by nightfall. It is better to leave a plant wilted (if not completely wilted down) in the afternoon and wait until the next morning to water it than to risk damping-off and other diseases by watering in late afternoon.

An organic mulch may be applied before or after seeding or transplanting, advises Larsen, and should be used in sufficient amounts so that when settled, the layer will be about two inches thick. The more finely divided mulches such as sawdust and rice hulls can be applied over the row ¼ inch in depth right after seeding, to aid germination by preventing crusting and drying of the soil.

Still another innovation is a project that involves growing 4 tomato plants around a circle of wire fence. Florida station horticulturists and the areas' newspapers refer to it as the Japanese tomato ring, although English gardeners have also been experimenting with it. Whatever it's called, though, and wherever used, harvests of up to 100 pounds or more of fruit are not uncommon.

The ring, which can be assembled in a weekend, calls for a length of wire fence (not chicken wire) 5 feet high and approximately 15 feet long; about two wheelbarrow loads of good soil; a layer of mulch or peat moss, some rich fertilizer, and the 4 young tomato plants. (A nematode killer is also listed, but organic gardeners won't want or need any chemicals.)

Locate the ring where it will get full sun and some protection from north and northwest winds. Clear a circle 7 feet across and dig the soil to a depth of several inches.

Arrange the fence in a circle, which will be about 5 feet in diameter, and place it in the center of your cleared ground. There should be a one-foot planting strip outside the fence. Place a layer of mulch about 6 inches deep in the ring. Add a layer of the good soil, another layer of mulch, and a final layer of soil. If you have no mulch, a 3-inch layer of peat moss can be used. A two-foot strip of screening placed around the bottom helps to keep mulch and soil in place.

Shape the top layer of soil to make a shallow dish to receive water and fertilizer worked into the soil in the ring. Set your 4 plants in the cleared space around the bottom of the ring. Top-dress with fertilizer and water the area around the ring when plants are small.

Tie plants to the wire with soft cloth or other material which will not damage stems. Vines will cover the wire and fill the inside of the ring. Be ready to pop up excess growth. Add fertilizer in the center of the ring about every 3 weeks, and keep the soil moist but not wet. A thorough watering once a week should be enough, but will depend on your soil, winds and other conditions.

—M. C. Goldman

Tomatoes Grown from Potatoes

Grafting tomatoes onto potato tubers—getting two major crops from a single growing plot—that's why we raise our own tomato-potato plants.

Start your tomato seeds in seed flats, just as you normally would. Then, when it is time to transplant, don't use a pot. Instead, set the young tomato seedlings in a seed potato with a one-inch round hole cut out of the center.

These tomato-potato "grafts" really take hold and produce double crops of juicy, firm tomatoes on healthily spreading vines and bushels of good-sized tubers growing in our compost-enriched soil. After placing the tomato roots in the potato hole, I carefully tamp soil around them, filling in so that the tomato stem is upright. Next I take a shallow cookie pan and put about one inch of soil into it, maybe

The potato-tomatoes are set in shallow cookie pans containing one inch of soil. Roots will develop and grow through into the soil while the "grafts" will transplant outdoors easily.

less, and lay the potatoes side-by-side in it with the tomato stems up. I have found that their roots develop and grow through the potato right into the soil where they take firm hold. This means that the young but vital root system will come up intact with the plant and a good earth ball when you are ready to transplant them outdoors.

Dig your planting holes in the vegetable patch about one foot deep and feed some compost into it, mixing it with a small amount of good garden soil.

The plants were grown with a host of other potash crops in a field that, a year or so before, showed a deficiency of potash when tested. The compost has taken care of that due to wood ash from limbs we burned right in the center of the field during the fall and plowed under. The experiment was a huge success, and next year we plan to grow more and experiment more.

I'm not out to claim that only this combined system will give you double crops while it withstands droughts, resists disease, and repels insects. The seeds you use—mine were Rutgers

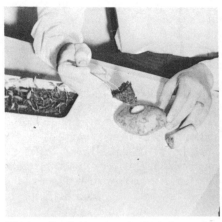

After inserting the tomato roots, firmly tamp soil so tomato stems stand upright.

—has a lot to do with it. But I do claim that *you will reap more per square yard of planting space than with any other major crop you may be growing!*

All you need is a little effort and time, a potato and a tomato seed, and a small hole in the ground. Combined, they'll give you two crops from a single small area—a harvesting fact worthy of notice both from the commercial farmer and the home gardener.

This year, try "grafting" your tomatoes on your potatoes.

—William A. Howes, Jr.

When To Mulch Tomatoes

I have learned this lesson: That if mulch is applied before the earth is thoroughly warmed, it will delay the ripening of tomatoes. I apply mulch now only when flowers are profuse, or may even wait until the fruit sets before mulching the plants. Then the mulch seals the heat in instead of sealing it out. Thus it pays to know when to mulch.

For late-ripening tomatoes I mulch my plants heavily when I set them

One-inch round holes are made in the center of each potato to take the seedlings.

out. For the earliest possible I set out enough to get ripe fruit in unmulched soil until the juicier and better-flavored tomatoes are ripened in the mulched rows. By the wise use of mulch you can prevent tomatoes ripening all at one time.　　—John Krill

Ding, Dong, Dell— Tomato in the Well

To get my tomatoes ready for the garden, I cut the tops off half-gallon milk containers, and slice around 3 sides to make a flap door. Round, gallon ice-cream cartons with slip-on lids are fine when you remove the bottoms, because they give the plants more growing room.

My sandy soil is workable almost as soon as the frost leaves, and I dig deep holes—my neighbors call them postholes — working a handful of sheep manure into each. After removing the lower leaves, I set the plants in the holes with the tops just below ground level, filling in with good, rich dirt up to the remaining leaves.

Then I slip a protective carton around each plant—again, check the picture—and bank soil all around the outside to the top. It is important to put the door hinge to the north and to tilt the carton southward to catch as much of the sun's warmth and light as possible.

The white sides of the cartons reflect enough light to keep the little plants happy even on gray days, and snug in their little wells they grow enthusiastically. I keep a little stone handy by each plant, to hold the flap door open in the day and on warmer nights, shutting it when night temperatures fall below 45. Frost may crust the outside, but the plants are dewy and safe inside. Before long the tomatoes are reaching out the tops, so I gradually lift the boxes, pinching off lower leaves if necessary, and letting earth crumble in below. By the time it is safe to remove the boxes, the well is filled and the plants are hilled. The long underground stems produce dozens of extra roots to supply the big, healthy plants which result from this method.

All my plants yield well. One cherry tomato plant covered a circle 12 feet

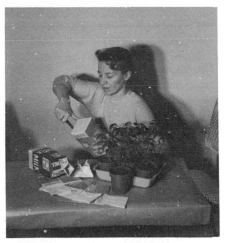

A cut milk carton makes a protective well for the out-of-doors tomato in the garden.

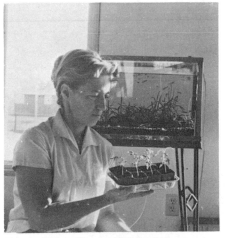

Marraine Miller with a pan of transplants.

Moving from the two- to the 4-inch pots. Hereafter, the plants will grow considerably faster and take up much more space.

3 sizes of tomatoes grown by the author.

in diameter on Labor Day, was loaded with fruit and still growing. It should have been staked, as the growth was so dense it was hard to get at all the fruit. Visiting in the blueberry patch, it tried to climb a neighboring dwarf apple tree, and overwhelmed a row of garlic. No insects have bothered my tomatoes—they're too healthy.

—Marraine Miller

Electroculture for Tomatoes
Magnets

Two horticulturists from the State University in Utah. Dr. Boe and Dr. Salunkhe, have shown that green tomatoes put under the South pole of a bar magnet will ripen much faster than control plants a few feet away. The study of biomagnetism and radiation has a direct application to all forms of life on earth. Biologists and horticulturists are beginning to study systematically the effects of natural and artificial magnetism on plants, seeds and trees. They coined the term "magnetotropic" in order to describe the sensitivity of plant life to magnetism and magnetic fields.

The Russian scientists Drs. Krylow and Tarakanova demonstrated in 1960 that seeds which are placed in the ground with the tip turned toward the South germinates long before the others. The buds of wheat, maize and pea seeds were arranged so that they faced either the North or South magnetic poles of the earth. The vigor of the growing plants whose seed buds faced the South magnetic pole demonstrated the validity both of the experiment as well as certain claims made many centuries back. The experiments were repeated with artificial magnetic fields and showed identical results.

And Cans

After reading about electroculture in OGF, I want to share with you my tomato planting idea.

The last of May I dig a hole in the garden about 3 feet across and 8 or 10 inches deep. Then I fill with about 3 inches of compost, cover with a little soil, and on this I set the tomato plant. About 3 inches from the plant, I put a double can. (I cut the top and bottom from two 46-ounce juice cans and solder the two together so they set high above ground.) This double can is set so both plant and can are in the ground about 6 inches. Then I fill the hole to ground level, which sets the plant deep. Also, if the plants call for water, it takes less water to get to the roots. I use the same method on peppers, cucumbers and melons.

Now to tell you how this improves the garden. My tomatoes are set 5 feet each way and cover the ground all over. They fruit earlier, and when there's dry weather I have to carry less water to keep them growing.

I thought it was just the water, but after reading your articles on electricity, I am thinking about that. I have had better tomatoes than my neighbors on the same kind of soil.

—Ralph A. Lyon

Antifreeze Cans

For more years than I can remember, I have used the one-gallon antifreeze cans (top and bottom cut out of can) in my garden. These are set over tomato plants—pushed into the ground just deep enough to prevent blowing away. The cans protect the plants from the wind and cutworms also. (I use only liquid manure for fertilizer and this makes it easier to measure quantity.)

A stake is driven outside the can on the east side. This explains to my satisfaction why I have no trouble raising pound or more tomatoes—the tin acts as electrodes.

I also use the quart-size cans around the lettuce plants. These I cut in half. Lettuce so grown is way ahead of open planting and makes it easy to fertilize them also.

I am glad to know why I have had such success.

As soon as the ground is warm, everything is mulched heavily. No weeds, and very little watering needed even in very dry weather. This year, will do the same thing with the melons, and will see if it improves growth and flavor as it does in tomatoes and lettuce.

—T. C. Goss

Grow Tomato Plants Indoors

Anyone with a properly mulched, weedless garden can obtain a stocky plant by cutting a branch off a bearing tomato plant, placing it in a container of water, and leaving it till roots appear, which takes about a week. Or a branch still attached to the plant can be placed on the ground, with soil over a 3-inch section, and left for 10 days or so. After roots grow, the branch can be cut from the larger plant and potted in a mixture of half compost and half light garden soil. We

To create a stocky tomato plant, cut one branch from a bearing plant, and leave it in a container of water till roots appear.

After the roots grow, the new plants may be potted in a 50-50 mixture of compost and light garden soil. Given a southern exposure, they will blossom in November.

Since the root systems of house plants are limited, it is necessary to prune vigorously to keep the tops in balance.

leave our potted tomatoes out in the garden for a few days, then bring them in on a warm day. Have them indoors for a week or so before the heat is turned on.

Tomatoes make very attractive house plants. We keep ours in a south window with flowering plants. They usually start blossoming in November and continue to bloom and bear tomatoes during the winter and spring.

Blossoming and bearing tomatoes are heavy feeders. Ours require bone meal, mixed in near the surface of the soil during February and March. We water them with rain water, compost water, or water that has stood at room temperature for several hours. Before they become too ungainly to move, we put them in the sink and shower them once a week, or even put them outdoors during unseasonably warm winter rains.

When they show a tendency to grow too tall we pinch the new growth or prune the plants. It's wise to remember that as house plants their root systems are quite limited com-

While indoor tomatoes may not bear heavily, the fruit is generally firm, and has good flavor and color —a winter treat.

pared to growth made in the outdoor garden. To compensate for this, the tops should be pruned; by balancing stem and leaf growth, setting of fruit is encouraged and heavier bearing results. As the fruit becomes heavier the plants need support. We tie them to a stake or to the window hardware.

To encourage pollination, simply tapping the plants firmly so that the pollen scatters is generally enough. Do this several times as new blossoms appear.

Our tomatoes did not bear very heavily, but the fruit was firm and had good flavor and deep color—quite a contrast to the pale, tasteless tomatoes sold in winter markets. We pot Doublerich (the vitamin C is double that of standard varieties) and sometimes Red Cherry, also high in vitamin C and high in sugar. Because of their delicious flavor, both are excellent in salads, and the attractive fruit of Red Cherry enhances the beauty of any fruit bowl. Cheerful yellow flowers, bright red tomatoes, and the green foliage of window plants show strikingly against the snow drifts and cold skies just beyond the glass.

—Devon Reay

Transplanting Technique

I have a method of transplanting tomato plants that may be of interest to some of your readers.

First I get individually potted plants at our local greenhouse, selecting those with the longest stems available. My main reason for this is to cut down on root damage and give the plants as little setback as possible.

I always take a chance and put two plants in the ground a week or 10 days before our last expected frost, which falls on Memorial Day. The rest go in as soon after that date as possible.

I usually have vine-ripe tomatoes around the first full week in July.

When planting, I dig a hole long enough to lay the plant in. The next step is to layer it with leaves—and don't spare them. Then prune off all but the top 3 leaves, lay the plant in the hole on its side, turn up the remaining leaves, and cover with loose soil.

My method accomplishes two things. First, it puts most of the plant below the surface and I need not worry about or do anything more to prevent wind damage. Secondly, the portion of the stalk that is placed below the ground will send out roots and give it just that much more of a jump.

—William E. Hathaway

Transplanting Helps Pollinate Tomatoes and Corn

Transplanting—moving tomato plants when in bloom from the starting bed to their outdoor rows—helps pollinate them.

That's been our experience out here in Phoenix, Arizona, where the nights are chilly and the soil is slow to heat up. We believe that the vibration caused by moving the plants 100 feet from the sunny, sheltered starting bed to the garden causes the fruit to set better.

Last time we moved the tomato plants, they were well-branched and in bloom. When we shifted them in mid-afternoon, the temperature was about 80 degrees, and they were jarred quite a bit during their ride on a little cart. Since each plant had been dug up with an 8½-inch ball of earth, they made quite a load.

We believe it's a good idea to shake the plants as the blooms open to encourage fruit-set. Some growers even use a vibrator in the greenhouse, and I plan to try an old-fashioned door-bell this spring in an effort to get earlier tomatoes. We did the same thing last year with corn, moving the 8-inch plants from starting bed to the garden where they were set one foot apart. They suffered no shock and gave us very early and delicious corn on the cob.

—H. R. Rawson

Onion Spray Protects Tomato Plants

Harry Botten, an Iowa gardener, grinds up onions, mixes the juice and residue with water, then sloshes the pungent solution on his tomato vines. "This system works perfectly," says Botten, pointing proudly to his 650 plants. "The fruitworms and aphids won't touch my tomatoes." He had lost crops to these pests for years, and had tried many insecticide sprays before developing his onion-juice treatment—a system already popular with organic gardeners, and often varied with garlic, hot pepper, etc., going into the solution. Botten applies the mixture after every rainfall, says it takes about a half-bushel of onions to make the needed 40 gallons for a planting as big as his. A tomato specialist at Iowa State Univ., Jack L. Weigle, expressed interest in Botten's method and results over the last two seasons. Acknowledging that certain kinds of onions contain compounds which inhibit the growth of fungus—also an enemy of the tomato—Weigle said, "You really can't rule out anything until you've tested it. And we haven't tested this type of treatment."

I Gamble on Early Tomatoes

The threat of frost is no reason to avoid getting an early start on toma-toes. I've been doing it—and without the cumbersome frost protectors most gardeners use.

Last year, early in March, 4 varieties of tomatoes (Beefsteak, Valiant, More-ton Hybrid and Burpee's Big Early) were sown in pure sphagnum moss under the new Gro-Lux fluorescent light in a cellar where the temperature ranged only between 50 and 60 de-grees. In late March, the seedlings were transplanted to 3-inch pots and transferred to a cold frame in mid-April. On May 3, all the plants were transplanted to the open garden.

The week following transplanting, night time temperatures averaged about 45 degrees, with the day high about 80. From then until June, the high temperature was never above 75, and the average was closer to 65. But on the 24th of May, the thermometer registered 31 degrees, with a barely visible white frost. From then into June, the night lows remained above 50, but the daytime highs were still cool—mostly in the 70's. On July 4th and 5th, two new low records—47 and 49 degrees—were set; and on July 9 and 10, still two more—43 and 47 degrees—went into the record.

Throughout this period, the plants were given no protection of any kind. We enjoyed our first garden-fresh tomatoes (Burpee's Big Early and Valiant) on July 11. Moreton Hybrid ripened the first on July 15 and Beef-steak on July 20. The significance in this particular season is that the record-shattering ups and downs of the thermometer had absolutely no retarding effect on the vine growth, flowering, fruit setting, or ripening of one of the "tenderest" vegetables.

There must be some explanation for this successful violation of long-established growing rules and, based on my experience, I would like to venture the following:

First—Plants started early at cool temperatures are relatively hardier and able to withstand extremes in temperature;

Second—Abundant sunshine which is available to the early plants makes for vigorous, strong plants and hastens maturity;

Third—Plants are really not as tender as reported.

It is my belief that the performance of my plants opens the door to further experimenting. So I advise my gardening friends to defy long-established traditions and the so-called accepted growing "facts."

—Richard Roe

But if They're Too Early . . .

If you plant your tomato seeds too early, they may be ready for transplanting before the frost has left your area. The Cooperative Extension Service of the University of Alaska has the answer.

"Fortunately," reports the station, "tomatoes will form roots anywhere along the stem, and this gives us a chance to simply take the tall, leggy plants and bury them deeper in the soil.

"One trick is to take a gallon can or a number 10 can and cut out the bottom, then cut down the sides and tie it together again with strings. Place this on a board and put in a little soil. Now put your tall tomato plants deep in the can, allowing a few good healthy leaves to stick out. Tomato plants will grow in a can of this size for another month.

"Then take the whole board, can, plant, and all into the garden; dig a large hole and slip the whole unit into it; pull the soil back around the can,

then cut the string that is holding the can together, roll it back and the roots of the tomato will not be disturbed at all. Water thoroughly with starter solution (manure 'tea' or liquid fish fertilizer is fine), and you should have no trouble getting some real early tomatoes.

"If you do not have a chance to re-set the tomatoes again, don't worry too much. When it's time to transplant them, simply make a little trench long enough for the stem and lay the tomato plant on its side. Gently bend up two or three inches of the tip, and then bury all of the roots and stem as if it had been planted in an upright position. The tomato plants will usually take off and grow without any great setback."

Light Waves to Grow Tomatoes

Experiments using light to make tomatoes ripen quickly and completely are being conducted at Clemson University by A. L. Shewfelt and J. E. Halpin. Writing in the University's Ag Research Bulletin, they stress that "patterns of wave lengths may be used to improve the odor, flavor, texture and nutritive value of tomatoes harvested at a pre-ripe stage."

Noting that temperature is chiefly responsible for the ripening of tomatoes which turn red most satisfactorily at 70 to 75 degrees, Shewfelt and Halpin report that while light is not essential for the development of red color, the rate of ripening increases somewhat when fruits are exposed to sunlight or incandescent light.

Three types of fluorescent light were used in the ripening tests, conducted at a temperature of 72 degrees: Cool-White; wide-spectrum Gro-Lux; and Standard Gro-Lux. Tomatoes kept in darkness at 39 degrees remained green hard, while those kept in the dark at 72 degrees developed color slowly.

Best results were obtained with the Standard Gro-Lux lights which operated for 16 hours a day, producing tomatoes with "consistently higher color values, a more desirable texture, being firm and mellow throughout."

The wide-spectrum Gro-Lux was next best, producing "higher color values" within 3 days than tomatoes exposed for 7 days to the Cool-White unit. The researchers reported at this stage of experimenting that "the differences in the rate of color development were attributed to the quality (wave length) of the light applied"

since the total intensity of light radiation was the same. So, if you plan to install a battery of lights for indoor propagation, be sure to check light wave lengths of the unit you are considering.

However, you are cautioned by Associate Director Halpin to work as much as possible with direct sunlight because artificial light alone probably will not be sufficient to induce tomato plants to set fruit satisfactorily, while the "cost of producing fruit under these conditions will be high." It would be advisable to use such lights for other less demanding plants, and also as a supplement to sunlight.

—M. F. Franz

TURNIP

Two-Pound Turnips— and Tender

I was tired of stringy, small turnips, so I decided to try the following experiment:

First I spaded the soil deeply and well, removing all weeds and grass. Next I spread 6 inches of compost over the plot to be planted, and followed this with very generous amounts of an organic fertilizer known to me as slag. It is a by-product produced when iron ore is smelted, and is high in lime and some trace minerals. (Fish emulsion

or other mineral-rich organic fertilizers will do.) After this was done I took the garden rake and worked the compost and fertilizer into the soil.

I made the planting ridge 3 feet wide and 12 feet long—about the average size for one packet of turnip seeds. Using the handle of the hoe, I made a very shallow trench down each side of the ridge and sowed the seeds as evenly as possible. The variety planted was Purple-Top, about the most popular of the white-fleshed turnips.

After the seeds were sown, I firmed

the soil down well using the back of the hoe, and covered the ridge with a very light application of straw, being sure it was dry and not easily packed down. This mulch kept the tiny seeds from being washed out of the soil by the sprinkler or heavy rain, and the method gave good germination.

When the turnip plants reached 3 inches, I thinned them. When they were 6 inches high, the second thinning gave me a nice mess of greens for the table. At this time I mulched with a 3-inch layer of grass clippings saved for the purpose. About two weeks later I gave the turnips another 3-inch layer of mulch—their final mulch for the season. My work from then on consisted of turning on the sprinkler whenever the soil needed more moisture.

The turnips grew so rapidly it was hard to believe—and the quality was far above any I have ever grown. They were sweet, not hollow or stringy turnips. But it was their size that was the amazing thing. They averaged two pounds each at maturity and even kept their good texture and sweetness! This experiment proved beyond a doubt that large turnips can be tasty and tender when grown organically.
—P. W. Smith

Spring, Summer and Fall

Instead of limiting myself to a fall harvest, I make my first sowing of turnip seeds as early in the spring as the soil can be worked. Being one of the hardier vegetables, turnips do very well in cool weather. But the fact is that they, like others such as cabbage, beets, carrots and onions, will do equally well during the warm season—though in their case this

seems to be less generally known, or maybe "contrary to popular opinion."

Also in common with these other "cool-weather crops," turnips can be used at various stages of maturity, and those started even in very early spring can be used from golf-ball size right up to full size. I've found that most turnips of this crop will stay solid and sweet throughout the warm spring into summer and on through cool autumn and cold winter. Though some develop cracks in growing, these are easily cut away—so that a single planting will suffice for the entire year if preferred.

If you don't want to devote much space to turnips, or prefer trying some where other vegetables have already occupied the ground earlier, fewer can be sown for a first crop then a later one grown the usual way. Actually, a rather short row or so will supply a surprising amount of turnips anyway, enough for an average family—and our own includes a miniature dachshund, named Taffy because she's long and sweet, who is "crazy" about turnips.

As for later use, turnips will stand plenty of frost and freezing weather. By the time we are well-launched into winter here in eastern Ohio— the real calendar winter, that is— their quality in the open, in more northerly regions at least, is likely to have deteriorated quite a bit, though not beyond edibility. A loose winter mulch of some extra hay or similar mulching material helps protect them for a time. Beyond this, well-stored specimens can be used until a new early crop is ready along about April.

—Richard L. Hawk

WATERCRESS

Guess What? No Stream!

I have always been under the impression that it was necessary to live along a brook or stream with running water to grow watercress. How wrong can one be? With my easy method you can grow it in your own garden under your own complete control.

To begin, you will need a length of roofing paper or some other insulating material, such as plastic sheets. I used a piece of 90-pound slate-surfaced roofing paper about 14 feet long. For a growing medium I used peat moss, ground rock phosphate, and granite dust. For fertilizer and nitrogen I used fish emulsion fertilizer.

Make the bed by laying the roofing paper or sheeting out flat where you want it to be and make an outline of the paper in the soil. Remove and scoop out a 6-inch-deep basin, keeping it about 4 inches smaller all around the outline. Level the dirt around the edges of the basin and be sure the bottom is level. Fit the paper back into this basin with the edges turned up all around. Into this basin put the peat moss, phosphate rock and granite dust, then mix thoroughly. Sprinkle with water until the mixture is completely soaked and wet. I used 50 pounds of phosphate rock, 50 pounds of granite dust, and enough peat moss to make the mixture about two inches deep in the bed. Keep wet and allow to stand for about a week.

Now you are ready to plant the seed. Scatter the watercress seed over a small portion at one end of the bed and cover with a light sprinkling of granite dust. Keep moist and the seed will germinate in about 5 days. When the seedlings appear, keep raising the water level until the plants are growing in water.

When the plants are about two inches high, transplant to the rest of the bed, spacing them about 6 inches apart both ways. Once the plants are set, fertilize with a cup of fish emulsion, and repeat applying this from time to time during the growing season. In about two or three weeks the plants will be ready to harvest—and the more you cut them, the thicker they will grow. In case you want to start another bed, you can take cuttings from the old one.

Should you care to make a permanent growing bed, you can do it with concrete. Be sure that your bed is just wide enough so that you can reach to the center from either side.

—Paul E. Mahan

192

YAM

Give Your Yams the Slip

My husband decided to try growing his own yam slips. Clutching a big 5- by 9-inch specimen of his favorite vegetable, he started in. In a small wooden box he placed an inch of equal parts garden soil, sand and compost. After splitting the tuber from end to end, he laid the pieces cut-side down on the dirt, and covered them with equal parts garden soil and compost. He then dampened the whole thing and placed the box on a large piece of plastic on the kitchen cabinet near the range. The undercovering was simply to protect the cabinet top, but over the box he tucked in another piece of plastic to help hold moisture.

In just a few days the tiny sprouts which were in evidence at time of planting had reached the top and the cover was removed. We continued to keep the soil moist but not soggy-wet. We prepared a good ridge in the garden with compost mixed into it, and by mid-April we were setting plants — 6 to 8 or even 10 at a time as they accumulated in the seedling box — until the row was finished. And did those vines grow.

My husband put the sprinkler or soaker on them a couple of times during dry, hot summer days. But early fall rains made further irrigation unnecessary. By mid-August we were grabbling big delicious hearts of gold.

INDEX